FIGHTING SKILLS

OF THE

SAS

SIDGWICK & JACKSON
LONDON

FIGHTING SKILLS OF THE SAS

MIKE ROBINSON

First published 1991 by
Sidgwick & Jackson Limited
First paperback edition published 1991 by
Sidgwick & Jackson Limited
a division of Pan Macmillan Publishers Limited
Cavaye Place London SW10 9PG
and Basingstoke

Associated companies throughout the world

ISBN 0 283 06087 5 hardback

ISBN 0 283 06133 2 paperback

Quoted excerpts taken from *One Man's SAS* by Lofty Large,
published by William Kimber & Co. Limited (1987).

9 8 7 6 5 4 3

A CIP catalogue record for this book is available from
the British Library

Printed and bound in Great Britain by
BPCC Hazell Books Ltd, Aylesbury, Bucks
Member of BPCC Ltd

Editor: Peter Darman

Design: Brown Packaging Limited

Artwork: Tony Randell

Front cover: Jungle skills are an integral part
of SAS training.

CONTENTS

LT GEN SIR PETER DE LA BILLIERE

PREFACE

This book is an examination of the skills and techniques employed by the Special Air Service in their operations, be they behind enemy lines, as during the 1991 Gulf War, or storming buildings and aircraft to free hostages from armed terrorists, as at the siege of the Iranian Embassy in 1980. The number of chapters, each one covering a different skill, is testimony in itself to the multitude of skills present within the Regiment. The proceeding pages illustrate how those many different skills have been used in the Regiment's campaigns throughout the world.

The SAS has fought many battles and campaigns since they were formed in the desert of north Africa 50 years ago, the vast majority of which they have won. The reason why they have been victorious in many different environments and often against incredible odds is not because they are supermen, though their enemies certainly believe this to be the case. Nor are they special because of the high-tech equipment and weaponry at their disposal, though they do have recourse to such hardware. Rather, it is because of the thorough and wide-ranging training each SAS soldier receives in every aspect of warfare, coupled with the physical and mental determination to get the job done. For the cornerstone of their success has always been, and will always be, the individual SAS soldiers themselves. It is they who win, because they dare, often against impossible odds and at the cost of their lives. To these men this book is dedicated.

Mike Robinson

SAS veteran of Oman, Malaya, Aden and commander of the Regiment during the Falklands War — Sir Peter de la Billière. His illustrious career ended with him being appointed Commander-in-Chief of British forces during the 1991 Gulf War.

JOINING THE SAS

Passing Selection Training to join 22 SAS is extremely tough. But the Regiment does not aim to break men or shatter their morale. What it wants are recruits who can think straight and keep on going when other soldiers just want to lie down and die. To this end the Regiment has devised a very thorough sifting process, one that ensures that only the right men get through. It is punishing, some would say brutal, but it guarantees the Special Air Service Regiment remains the best elite force in the world.

Cold, wet, tired and hungry, the soldier forced himself on to further feats of physical endurance. Despite the punishing pace and the difficult terrain he had to cover, he was determined to finish the course within the time set by his SAS instructors. To make matters worse the weather had closed in, reducing visibility down to 20 metres or less. Driving rain stung his eyes, obscuring his vision as he tried to orientate his map. Where the hell was he? He looked once more at his compass, uncertain as to his location. Navigation over such featureless terrain relied solely on good compass work and dead-reckoning. Reasonably confident that he had not

The SAS look for both mental and physical toughness in their recruits, men who will push themselves to the limit and will still be able to think and shoot straight.

made any major cock-up in his navigation, the soldier limped off once more. As his blistered feet settled back into their routine, the pain in the muscles of his legs lessened marginally as his pace steadily increased back to a quick walk. His ultimate aim of passing Hereford's notorious Selection Course was forgotten for the time being as the soldier concentrated on making it to his next rendezvous.

Strange though it may seem, the mountains of south Wales were initially chosen as the venue for SAS Selection because of the similarities the area has to a jungle environment. At first glance one might be forgiven for wondering just what very bleak, windswept mountains, often covered with dense cloud and rain, might have in common with thick, impenetrable, humid jungles. To understand the answer one has to realise that the only way to navigate in the jungle is by the compass and distance method, and the same applies to the Brecon Beacons and the Black Mountains, especially when the weather closes in and visibility is reduced to a few metres. There are mountains in the jungle too, just as difficult to traverse. Both environments are hostile, although in different ways, and both test those who must survive in them to the limit.

From the beginning the SAS has comprised volunteers from all arms

Since the Regiment's reformation in the jungles of Malaya after World War II, the SAS method of selecting its soldiers has been a primary reason for its adaptability and, ultimately, its success. During the early stages of Britain's post-war campaign in the Far East, it was rapidly realised that counter-insurgency (COIN) operations deep in the jungle required a special type of soldier. From the beginning the SAS has comprised volunteers from all arms and corps of the British Army (as well as a few from the Royal Air Force); men with exceptional abilities and a high level of motivation.

Unlike a number of special forces (SF) units around the world, Britain's SAS Regiment does not accept direct-entry volunteers. All prospective SAS troopers must have previous service with another more 'conventional' formation before volunteering for the SAS, and only after successfully completing a tough selection process will they be allowed to start training with Britain's special forces. Fitness, endurance and mental preparedness are all essential qualities which must be possessed by the successful candidate. SAS Selection is a test which can only be passed by a soldier, or officer, who has thoroughly prepared himself for it.

The course is a severe test of an individual's mental and physical stamina

The SAS Selection and Continuation Training programme is an arduous, tiring and lengthy process, and only the toughest, fittest, most determined soldiers can survive it. The four-week Selection Course takes place twice a year at the Regiment's base at Stirling Lines, Hereford, and in the mountains of south Wales. The course is a severe test of an individual's mental and physical stamina and land navigation skills. As the weeks progress the marches get longer and the recovery time gets shorter. Many find the course too punishing and give up. The size of the groups conducting marches is gradually reduced, until individuals are making solo timed navigation exercises. These test an individual soldier to the limit of his endurance, and then beyond. The culmination of Selection is the 'long drag', a timed 60km march over the Brecon Beacons which all potential recruits must pass.

If he passes, a soldier will, after a week's leave, move on to Continuation Training. This lasts for around six months and tutors the potential recruit in all the SAS patrol skills: weapons handling, combat and survival, reconnaissance, medicine, demolitions, camouflage and concealment, resistance to interrogation, etc. Continuation culminates in the soldier being tested, along with three other hopefuls, in a four-man patrol jungle exercise in the Far East. The failure of any one man in the patrol will result in the other three also being 'binned'. Only at the end of all this training and testing, and on completion of a basic static-line parachute qualifying course, will a volunteer eventually be 'badged' SAS and attached to an SAS 'Sabre' Squadron.

Above: The toughening process continues. Potential troopers are put through their paces.

Britain's regular Special Air Service Regiment (22 SAS) has four 'Sabre', or fighting, Squadrons, each divided into four specialist troops of 16 men. These troops are further divided into four patrols, each comprising four men. On his return to Hereford the newly 'badged' SAS trooper is assigned to a troop, where he undertakes training in their particular specialisation. Within a 'Sabre' Squadron each troop has its own specialised role: Boat Troop (amphibious warfare); Mobility Troop (land vehicle operation specialists); Air Troop (military freefall parachuting); and Mountain Troop (mountain and winter warfare). In addition to his troop skill, each newly-qualified volunteer must acquire a personal or patrol skill, either medicine, demolitions, communications or a language. Each four-man patrol contains at least one man with each skill, although as their length of service increases men acquire additional patrol skills. This cross-training is one of the Regiment's great strengths.

Highly motivated, stringently selected and exceptionally well trained, a soldier serving with the SAS has undoubtedly reached the apex of professionalism. He is ready to be deployed to any 'hot spots' around the world; to be inserted deep behind enemy lines to undertake sabotage and assassination missions; to free hostages from the clutches of armed terrorists; and to do the seemingly impossible, to march on foot for hundreds of kilometres, to surmount any obstacle, and conquer the world's most hostile environments.

He will do these things not in the quest for glory, for the deeds of the SAS are rarely fully revealed, but because he is proud to be a member of the world's most elite fighting force. It is a strange fact that the discipline within the Regiment is surprisingly relaxed; there is no yelling or snapping of orders. Yet each SAS soldier has, within himself, an iron-hard self-discipline which is far stronger than any set of imposed regulations. This is what motivates him to perform seemingly superhuman tasks, often at the cost of his life. He is in every way the ultimate warrior.

WEAPONS SKILLS

Expertise in the handling of many different types of hand-held weapons, from handguns to sophisticated anti-aircraft systems, has enabled SAS teams, often hopelessly outnumbered, to achieve victory against well-armed and determined opposition. Constant intensive training in such environments as the Regiment's 'Killing House' ensures that the men of the Special Air Service are second to none in weapons handling. Their finely-honed skills are at their most devastating during a firefight, when an SAS team can lay down a murderous amount of accurate and sustained fire.

On 19 July 1972 over 250 heavily armed guerrillas of the Dhofar Liberation Front attacked nine SAS soldiers of the British Army Training Team stationed at the Omani town of Mirbat. Using accurate and sustained fire from their assault rifles, machine guns, a mortar and an old 25-pounder field gun, the SAS, supported by a handful of Omani gendarmes and tribesmen, successfully beat off repeated attacks. At the end of the battle 30 guerrillas lay dead on the battlefield, with many more withdrawing wounded. The SAS had just two men killed. The proficiency shown by each SAS trooper in the handling of different weapons, from assault rifles to a

A British Army 81mm mortar opens fire during a training exercise. This weapon was used with great success by the SAS during the 1982 Falklands conflict.

field gun, was responsible for this stunning victory.

SAS personnel have always used a wide variety of weapons, from heavy machine guns to handguns. They will have been taught weapons skills during their initial basic training, before they joined 22 SAS. After completing Selection the prospective troopers then receive intensive instruction in the use of many different types of small arms and support weapons. They learn the principles of weapons handling and employment, and how to get the most out of a particular firearm. The SAS ground rule is flexibility: different weapons suit different types of mission. And this means that their approach is often radically different from that of the rest of the Army. For example, the handgun has traditionally been a weapon of last resort. The SAS, however, often use it as a primary offensive weapon.

The High Power is a favourite within the SAS Regiment

The standard issue British semi-automatic pistol is the Browning, FN 9mm High Power, No.2 Mark I, although it is more often referred to more simply as the Browning High Power. Designed and developed by *Fabrique Nationale* (FN) of Belgium, a company also behind the self-loading rifle (SLR), the 9mm calibre Browning pistol replaced the Webley revolver previously in widespread service with all arms of the British services.

The High Power is an acknowledged favourite within the SAS Regiment. The advantages of the semi-automatic pistol over the revolver are essentially the number of rounds available to the firer, the ease with which the weapon can be reloaded and the lighter weight for an equivalent barrel length. In addition, the High Power is magazine-fed and takes 13 rounds (14 with one in the chamber), while a revolver works on the cylindrical chamber principle and normally accepts a maximum of six rounds. The single advantage of a revolver is that should a round mis-fire, i.e. fail to detonate, a semi-automatic pistol will have to be re-cocked manually while a revolver requires its operator simply to squeeze the trigger one more time. For the above reasons revolvers are seldom, if ever, used by members of Britain's special forces (SF).

High Powers have been used by the SAS on a number of operations. During the Aden campaign (1965-67), for example, SAS troopers were formed into undercover teams called 'keenie-meenie'. Mingling with the local population, they became expert at quickly drawing a High Power from the folds of an Arab *futah*. At the Iranian Embassy siege in 1980 High Powers were carried as back-up weapons to the troopers' Heckler & Koch submachine guns. At that time SAS members undertaking undercover operations in Northern Ireland were also using the smaller 9mm Walther PPK handgun.

For many soldiers who are selected for training with 22 SAS, the first time they will come across the 9mm Browning handgun is after the Continuation phase of their training. Instruction in this type of weapon therefore begins with the basics. Students are introduced to the weapon, taught how to strip and assemble it, methods of loading and unloading, and what safety procedures are applicable in its use.

The pistol is an excellent weapon for engaging targets of opportunity at close quarters (50m or less), and its short barrel enables the soldier to rapidly engage suddenly-exposed targets with either hand. But because its barrel is short, the need to aim accurately is of paramount importance. In untrained hands the Browning, like any pistol, can endanger the life or physical well-being of the user or comrades-in-arms alike.

The pistol is an excellent weapon for engaging targets at close quarters

Intensive training in the use of the High Power is therefore given to all SAS personnel. Students are instructed in the double-handed alert and firing positions, where the weapon is held in both hands and brought up in line with the centre of the body. Accurate pistol shooting takes time to perfect, therefore SAS students at Stirling Lines are given plenty of opportunity to practise their pistol shooting on the 30m range in Hereford or, if they are selected to join an SAS Counter Revolutionary Warfare (CRW)

The Browning High Power, a handgun noted for its reliability, durability and relative accuracy.

team, they will become familiar with firing their High Powers inside the 'Killing House', as the Close Quarter Battle (CQB) range at Hereford is called. They are taught the following pistol firing techniques:

1) When engaging targets of opportunity at ranges of 10m or less, the alert position should be adopted, i.e. square on to the target, and double taps (two rounds fired in quick succession) fired in rapid sequence.

2) The pistol should be held firmly and pressure applied equally between the thumb and fingers of the firing hand.

3) Breathing should be restrained (as when firing the rifle) and the 'squeeze' on the trigger should be maintained while the round is fired and immediately afterwards.

4) Release trigger hand pressure once the shot has been fired and as the recoil diminishes, and allow the pistol to settle back onto target.

5) Watch the rounds strike the target and make any adjustments to your aim accordingly.

6) If the target is between 10 to 20 metres away the aim should be raised.

7) When firing the pistol count off the number of rounds left in the magazine.

8) When necessary reload with a full magazine even though the used one may have some rounds unfired.

Above: When firing a handgun two-handed the arms are locked and the legs slightly bent.

9) Having dealt with the enemy, and when no further shooting is likely, make the weapon safe.

These techniques can be used with any pistol, not just the High Power, and can be adapted for different types of operation such as undercover operations or firing around cover, where a single-handed stance may be better suited. The SAS Regiment has a wide range of semi-automatic pistols in its armoury, and can choose the right tool for the job. Depending on the particular mission, SAS soldiers might carry a Browning 9mm or one of the new generation semi-automatic pistols such as the VP-70 from Heckler & Koch. This weapon can fire either single shots or three round bursts, making it an excellent close quarter weapon in the hands of a highly trained SF soldier.

Other SAS favourites are the Ingram Model 10 and Model 11 submachine pistols; weapons that bridge the gap between pistol and submachine gun (SMG). Designed by Gordon Ingram with a view to effectively arming downed aircrew, and manufactured in the USA, these weapons have proved a popular choice for SF operations. Short and easily concealable, they range from 9mm to .45in in calibre, have an extremely high rate of fire of 1150 rounds per minute (rpm), and have been found to be especially suited to undercover operations such as those conducted by the Regiment in Northern Ireland.

The submachine gun is another, more traditional, close quarter weapon used by the

Below: Firing a handgun one-handed. The firing arm is bent to create a shock absorber effect.

SAS. The standard issue weapon of this type is the rather aged L2A3 Sterling. A replacement for the ubiquitous Sten gun, it chambers a 9mm calibre cartridge. It has a much greater offensive capability than the Browning pistol, is capable of fully-automatic fire, has a magazine capacity of 34 rounds, and has a cyclic rate of fire of 550rpm.

Though useful for street fighting, house clearing and general close patrol work, the L2A3 Sterling, because it makes so much noise when firing, has only limited use as an SF weapon. However, within the same family there is another version designed specifically to meet the SAS requirement for killing enemy sentries or for conducting ambushes, tasks where silence is of primary importance. The L34A1 or 'Silenced Sterling' is a specialist weapon. It is fitted with an integral silencer which, by reducing the speed with which the bullet and accompanying gases leave the barrel, almost completely eliminates the noise of the explosion.

The L34A1 is essentially a modified L2A3. It fires the same ammunition, but has a differ-

Below: CRW training. The High Power's 13-round magazine gives the weapon excellent sustained fire capability, making it ideal for hostage-rescue work and general undercover operations.

Because they are designed for concealment and a quick draw, crossdraw holsters are favoured by SAS soldiers when operating in plain clothes.

ent back-sight to allow a greater field of vision for use in low-light conditions and for close-quarter work. It should not be fired in the burst mode as this reduces the efficiency of the silencer, and will on occasion leave a smoke 'signature' should the silencer have a trace of surplus oil.

Over recent years 22 SAS Regiment has developed and expanded its CRW Wing and as a result has acquired considerable expertise in the testing of new items of equipment and weaponry. As leaders in the art of counter-terrorist warfare, the SAS has, since the 1960s, operated a number of exchange programmes with American and European SF units. This has led to the exchange of information and the adoption by the SAS of new weapon models and designs.

One such design is the MP5 series submachine gun. Manufactured by Heckler & Koch, Germany's leading small arms firm, these 9mm weapons were used by 'Pagoda' Troop during the Iranian Embassy siege-busting operation in London. First introduced in the 1960s, but continually modified according to 'user experience', the MP5 series SMGs enjoy a reputation of being both sophisticated and reliable.

Fitted with a 30-round magazine, the MP5 used by the SAS has a higher rate of fire than the L2A3 — 800rpm as opposed to 540rpm — which means that a trooper can get more shots into a target in a shorter space of time. Another useful feature of the MP5 series is that they can fire three-round bursts, as well as single shots or full automatic fire, giving the the ability to 'treble tap' with a single squeeze of the trigger.

The MP5 used by the SAS has a higher rate of fire than the L2A3

Heckler & Koch produce a number of impressive accessories for their MP5 series. These include a .22 calibre conversion kit, useful for indoor live-fire training; a blank firing adapter; a muzzle-mounted grenade-launching attachment; and a variety of optical devices ranging from night scopes to aiming point projectors which put a red dot on the target.

Other variants of the MP5 presently in service with the SAS include the MP5K, the last letter meaning *kurz* or short. This weapon is a mere 325mm in length, as opposed to the 490mm of the MP5A3, making it especially suitable for CRW and clandestine operations. With a 10- or 15-round magazine, the MP5K is an effective replacement for the pistol. In addition, there is the MP5SD range. It is a silenced series of SMGs which make the SAS trooper less obvious to the enemy, by reducing both noise and muzzle flash. When used with an aiming point projector the MP5SD can be visually sighted on to a target by moving the tell-tale red dot, which indicates the mean point of impact (MPI). This means that the operator does not need to bring his weapon into the shoulder or even adopt a conventional firing position, thus allowing him a far greater degree of tactical flexibility while virtually guaranteeing a first-round hit.

Although firing the SMG from the prone (lying), sitting, and kneeling positions are taught to special forces, the two most common firing positions are either from the shoulder or the waist while standing.

When firing from the shoulder the soldier discharges an aimed shot or a number of single aimed shots. If time is too short for proper aim the SMG can be fired by sense of direction. The aiming point should be the central area of the target and bursts should only be fired at groups of the enemy at very close range. Firing rapidly and accurately with the SMG while standing is a matter of balance and footwork, and the correct posture, feet astride and leaning slightly forward, can be achieved by regular practice.

Firing the SMG while standing is a matter of balance and footwork

Firing from the waist is the method employed when the enemy is met at point-blank range, if there is no time to raise the weapon into the shoulder, or the SMG's butt is folded or has not been extended. In extreme close quarter situations the weapon can be held centrally in the middle of the stomach with the head right over the barrel, with the firer jumping around to engage targets square-on. While these positions are peculiar to the SMG, most other positions are the same, or similar to, those adopted when firing the assault rifle.

Adopted by the British Army in 1956, the Stirling submachine gun is extremely reliable. The silenced version is used by the SAS in Northern Ireland.

Sniper rifles, on the other hand, are true specialist weapons requiring particular skills. Designed to meet a specific operational requirement, in the hands of a skilled marksman they are one of the most deadly and selective weapons on the battlefield. Using specially accurised rifles with telescopic sights, trained snipers are the most feared of almost any type of soldier. Each year a small number of specially selected SAS troopers attend sniper courses, which last from four to six weeks depending on which course they attend, to be instructed in the skills needed to hit targets at great distances with one round, and remain undetected.

Below: When firing an MP5A3 with extended stock from the shoulder the weight is on the leading leg, with the body leaning into the gun.

Special forces around the world employ snipers in three main roles: in the field (conventional or counter-insurgency warfare); in the urban environment; or for specific CRW missions.

The sniper, operating either alone or with a partner, can pin down a large enemy force by killing its leaders and communications specialists. These targets are often identified by means of their equipment, such as binoculars, radio sets and antennae. Because they usually work closely together they can often be picked off with two single shots, fired close together, thereby denying the enemy their command and control capability in the space of two or three seconds. Other primary sniper targets include machine gunners; other troops manning crew-served weapons, such as anti-armour missile launchers and air defence weapon systems; and electrical and mechanical equipment.

The sniper can pin down a large enemy force by killing its leaders

Good eyesight and ice-cool nerves are among the primary qualities considered when selecting a man for sniper training, but there are many more. Marksmanship ability itself is not enough; a high level of camouflage and concealment skills, plus advanced fieldcraft, are vital to the sniper if he is to be effective on the battlefield. Snipers must also be able to move across open terrain without being spotted by the enemy, select concealed firing positions which offer a good view of the target, and have an avenue of escape once the target, or targets, have been effectively engaged.

All SAS soldiers are taught how to maintain their weapons, but for a trained sniper the standards are far higher. Too much oil in the weapon, for example, will create a tell-tale

Above right: Use of cover presents less of a target to the enemy when firing a submachine gun from the prone position.
Right: The MP5 series of submachine guns are widely used by the SAS in their hostage-rescue operations. They have four firing mode settings: safe, full automatic, three-round burst and semi-automatic.

smoke 'signature', giving away the sniper's position to the enemy. This is also true of the sun glinting off a shiny barrel or any other part of the rifle. In addition, sniper weapons are generally less robust than those issued to other troops, and the sights in particular must be treated with great care. Together with specially selected high quality match ammunition and technical expertise, the high resolution optical sights give the sniper his advantage. Without them the sniper loses his ability to hit targets at great distances, and is little better off than a lone infantryman.

The individual model used by the SAS is the Accuracy International PM

The range of sniper rifles varies according to the manufacturer, the model being used and the type of sights fitted, but most are accurate up to ranges of between 900-1000m. Until 1987 the British Army's sniper rifle had been the venerable .303in Lee Enfield. The No.4 model rifle had been replaced as the standard infantry weapon in the late 1950s by the L1A1 SLR, but the No.4 was retained as the official sniper rifle in the guise of the L42A1, a Lee Enfield No.4 Mk (T) .303 calibre converted to 7.62mm calibre. In the mid-1980s, trials for new sniper rifles were concluded and the Lee Enfield replaced.

The individual model now used by the SAS is the Accuracy International PM, which can fire 7.62mm calibre ammunition. The PM, designated the L96A1, is a bolt-action weapon which fires ammunition fed either from a detachable 10-round box-type magazine, or fed singly through the ejection port when the bolt is pulled back. Its butt and stock are manufactured from high-density green plastic, and it has an adjustable rubber butt-pad, an important consideration bearing in mind the length of time a sniper might have to spend motionless before firing. In addition, the rifle has a bipod and a retractable spike on the rear of the butt, so the weapon can rest comfortably during long periods of observation.

The L96A1 comes with a Schmidt & Bender PM6 x 42 telescopic sight, chosen for its high degree of ruggedness, excellent clarity, and light transmission of 93%, which makes it especially useful for shooting in low-light conditions. On a recent snipers course run by the Royal Marines at their own Commando Training Centre in Devon (a course popular with the SAS, Western European and US special forces), all 13 students using the system achieved the highest possible score at ranges of between 300-600m. The fct that these results were achieved under realistic operational conditions and say much for both the standard of training and the weapon. The SAS have, in the L96A1, an extremely accurate, robust, high-quality weapon, specially tailored to meet their needs and worthy of their skill in this extremely demanding area.

Above: The underarm assault position. Weight is on the leading leg and the body leans into the weapon; the sling gives support when firing.

While sniper rifles are reserved for a few selected specialists, the ubiquitous assault rifle is the standard-issue personal weapon for the vast majority of soldiers, whether they belong to special forces, infantry or supporting arms and services. The assault or battle rifle is the basic tool of almost all ground troops. The difference between these two types of infantry weapon is marginal, the assault rifle having a fully-automatic as well as semi-automatic capability, while the battle rifle may be a semi-automatic weapon only. Both types of weapon fire a more powerful and heavier round than the SMG, either the new NATO SS 109 5.56mm round or the older NATO 7.62mm, as opposed to 9mm, have longer barrels, are more accurate and have greater range. Since the late 1950s the SAS has mostly used the SLR. This 7.62mm Belgian-designed semi-automatic rifle is now being replaced in British Army service by the British-designed and manufactured bullpup 5.56mm SA-80.

Below: Firing an Ingram submachine pistol one-handed. The elbow is locked and the sling is grasped by the forefinger for greater control.

Although the SA-80 began to enter service from the late 1980s onwards, almost all SAS soldiers received their initial training on the SLR. As most men who volunteer for service with the SAS have already served five or more years with another Regiment or Corps (the minimum requirement is three years), it is likely that for the majority the SLR represents their first individual weapon.

The SLR is an excellent soldier's weapon; reliable with very good stopping power, i.e. the weight of the round hitting the target is heavier in relation to smaller calibre weapons. It fires standard NATO 7.62mm ammunition, a heavier but more effective round than the 5.56mm used in the American M16 automatic rifle and the SA-80. The weapon itself is robust or 'soldier-proof', and is accurate well beyond the sighted range of 600m.

The SAS has used the SLR everywhere, from the arctic wastelands of Norway, the steamy jungles and swamps of Southeast Asia and Central America, to the deserts and mountains of the Middle East and the Arabian Gulf. It has proven itself operationally in Malaya, Borneo, the Radfan, Oman, as well as more recently in the Falklands, where it armed SAS, Royal Marines Special Boat Squadron (SBS), and Mountain and Arctic Warfare (M&AW) Cadre patrols.

The SLR is an excellent soldier's weapon with good stopping power

As well as the SLR, the SAS has also used the M16 for many years. The former is a heavier weapon, weighing 5.25kg with a 20-round magazine, while the M16 weighs 3.72kg with a 30-round magazine. The two rifles complement each other — the SLR with its longer range and heavier punch; the lightweight M16 capable of full automatic fire and being able to put down a considerable amount of firepower in a short space of time. In addition, the M16's lighter weight means that more ammunition can be carried by each patrol member, an important consideration when a unit is operat-

Above: The state-of-the-art Accuracy International Model PM Sniper Rifle (L96A1).

ing unsupported in a hostile environment for any period of time. British SAS patrols in Borneo used a mixture of both weapons, as did the Australian SAS when they deployed teams in Vietnam, and the results in combat matched the expectations and high standards of those relying on them.

The first M16-type weapon employed by the SAS Regiment was the Armalite AR-15, developed for the US Air Force and issued to American Green Berets operating in Southeast Asia. When the US Army adopted the weapon to arm its troops in Vietnam, the manufacturers changed from being Armalite to Colt, and the designation switched from AR-15 to M16. The SAS used the Armalite during their campaign in Borneo in the early 1960s, and the designation 'Armalite' has remained in SAS usage ever since.

Other major benefits of the M16 include its diverse range of attachments: a bayonet for close quarter combat; bipod, for accuracy when firing from the prone position over longer ranges; as well as telescopic sights and night vision aids. But the M16's most important attribute is that it can be fitted with an M203 40mm grenade launcher. This weapon, adopted in relatively small numbers by the British Army, is a favourite with units such as the SAS. It is attached beneath the stock extending under the M16's barrel, with a separate trigger mechanism forward of the M16's magazine. The M203 fires a variety of 40mm grenades including smoke, high explosive (HE) and phosphorous, to a maximum accurate range of just under 400m.

The SAS used the Armalite during their campaign in Borneo in the 1960s

The latest version of the M16 is the M16A2, which differs from the M16A1 in a number of ways. It is slightly heavier than the earlier weapon, and has a much thicker profile. Other changes include a three-round burst capability, improved sights and strengthened non-metallic parts such as the butt, stock and pistol grip. It also fires the improved NATO 5.56mm ammunition, making the M16A2 a truly effective and accurate weapon in the hands of an expert.

Another weapon in the Armalite/M16 series that is currently used by the SAS, especially in Northern Ireland, is the versatile Colt Commando, essentially a much shorter, handier

Above right: Body well behind the rifle, sling high on the left arm, butt close to the neck — the classic sniper stance.
Right: SAS snipers are highly practised in the art of camouflage and concealment.

HIGH-VALUE TARGETS

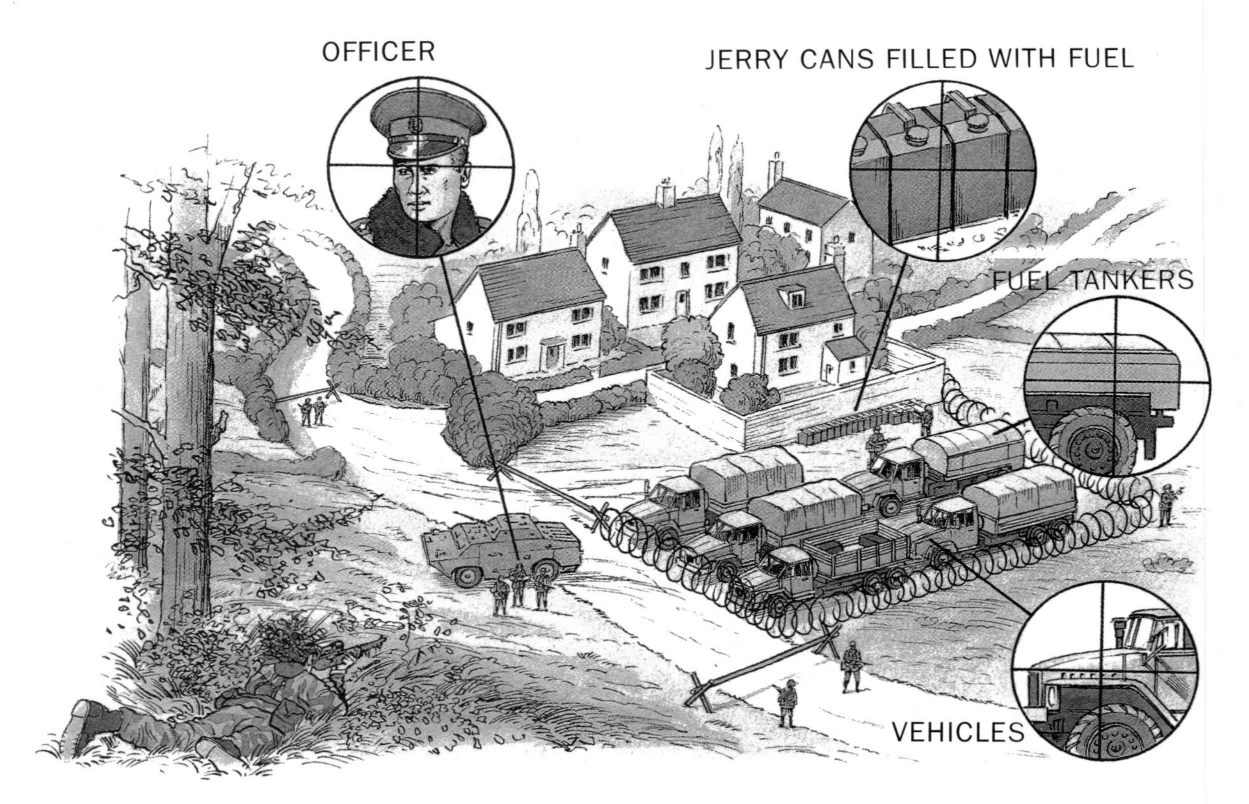

Above: Deep inside enemy territory, an SAS sniper, unseen by the troops below, selects various targets through his sights.

version of the AR-15. Designed as an aircrew survival weapon, it is a favourite with SF units. With a shorter barrel reducing its overall length to 787mm (butt extended) compared to the M16's 1006mm, the Colt Commando is ideal for close quarter and jungle combat.

Those serving with the Special Air Service Regiment find that their weapons skills are constantly being tested, not only with their own pistols, submachine guns, assault rifles and machine guns, but those of foreign armies as well. There may well be occasions during an SAS soldier's military career when his skill in the use of enemy weapons, or those of his allies, will be put to the test.

All SAS troopers are introduced to the basics of many foreign weapons during the Continuation phase of their training, which takes place after they have passed through the initial SAS Selection process. These weapons include all those manufactured by the Soviet Union and Warsaw Pact countries, most notably the AK-74, AK-47 and AKM assault rifles, the SVD (Dragunov) sniper rifle, and older weapons like the SKS carbine and the PPSh submachine gun. Students are also introduced to many other allied and NATO infantry weapons such as Germany's Heckler & Koch G3, Italy's Beretta AR70 and France's FAMAS.

As the SAS are essentially ground troops they operate mostly on foot, or occasionally mounted on small cross-country vehicles. Since most of the SAS soldier's equipment must be carried into battle by the man himself it must be relatively lightweight. The machine gun is therefore the most common 'heavy' weapon carried operationally. There are three basic types of 'gun', the general term used to describe

these much heavier, man-packed, section close-support weapons: the light, medium and heavy machine gun. All three types are used by the SAS.

In modern warfare there are two main roles in which the machine gun is used in support of ground troops: in the sustained fire role, providing defensive and offensive fire; and in the section support role.

Following the adoption of 7.62mm calibre ammunition as the standard NATO round, the opportunity was taken to select a modern general purpose machine gun (GPMG). After thorough trials the Belgian FN MAG was accepted for service with the British Army. The design incorporated features of the Browning Automatic Rifle (the working parts) and the German MG42 (the feed mechanism), both of which had proved themselves to be exceptional weapons.

Most soldiers will be familiar with the

Above: These two SAS soldiers on a field exercise illustrate the different types of assault rifles used by the Regiment. They carry the SLR and M16 respectively.

GPMG long they begin SAS Continuation training, but they will receive further instruction in this and other weapons at the nearby Senneybridge Training Area, and elsewhere in the UK and abroad. During training troopers have the opportunity to get to know the GPMG well. It accompanies them on runs over battle training areas such as Senneybridge, and punishing cross-country marches over the Black Mountains and Brecon Beacons of south Wales.

The L7A2 is a belt-fed weapon weighing 10.9kg. With its integral bipod it has an accurate range of 800m and when attached to a tripod (in the sustained fire role), it can fire 800 rounds a minute to a range of up to 1400m. The GPMG uses a light barrel in the light

Right: Because the M16 (shown here) has a light recoil, the firer can adopt a more erect stance than is normal when firing assault rifles.

machine gun role (a spare one is carried by the gunner). When the weapon is employed in the sustained fire role the butt assembly is easily removed and replaced with a back-plate. A dial sight, the same as that used on the standard 81mm mortar, can be attached for indirect fire.

The GPMG can provide an SAS patrol with the extra punch it needs

The GPMG can provide an SAS patrol with the extra punch it needs to carry out a successful ambush or to shoot its way out on a contact with a superior enemy force. Used in conjunction with other weapons, such as the SLR, SA-80 or M16, it can give a four- or eight-man patrol an advantage over enemy units of similar size, or larger groups who may not have the benefit of a heavy close-support weapon.

SAS soldiers learn how to fire the L7A2 from a variety of different positions. Like other ground troops they are instructed in firing the weapon from the prone position but, unlike conventional infantry forces, where two men serve the gun, one firing and one loading, the small size of the SAS patrol means that the gunner often does both jobs. In addition, troopers are also taught how to fire the weapon from the standing position with the bipod folded underneath the gun, usually from the hip but occasionally from the shoulder. When firing from the hip the soldier usually uses the weapon's sling to maintain the gun at waist height, and leans forward to counter the recoil.

The second method of firing a GPMG from the standing position requires a good deal of physical strength and involves the weapon being raised into the shoulder in the same way as a rifle. The ammunition belt is draped over the left arm, while the right supports the weight of the weapon. Only short bursts of up to four rounds or so can be fired in this manner as the

Right: The M16 has a high muzzle velocity, which means that any enemy personnel hit at ranges of under 200 metres are cut to pieces.

gun is difficult to control, but it can be used effectively for suppressive fire at short ranges.

The GPMG is a popular weapon despite its weight and that of its ammunition. The 7.62mm dis-integrating link ammunition used by the L7A2 is distributed throughout the patrol, lessening the weight to be carried by the weapon's gunner in the same way that the patrol signaller's spare batteries are spread out among his colleagues. Though production of the GPMG has now been completed, it is likely to remain in the SAS Regiment's arsenal for some time to come.

While the L7A2 is capable of providing heavy accurate fire over great distances, the SAS uses another type of weapon capable of causing the same sort of devastation, but at short ranges. Initially introduced during the Malayan campaign, the shotgun provided the SAS with a formidable weapon for close quarter work in the thick, impenetrable jungle. Early models were standard civilian shotguns or those developed from them for police riot-control operations. Loaded with buckshot rather than the non-lethal birdshot used by riot police, shotguns were often issued to a patrol's lead scout, enabling him to put down a spread of deadly projectiles with a single shot.

Shotguns were often issued to a patrol's lead scout

Different types of shotgun ammunition can be carried, depending on the mission and the type of threat likely to be encountered. Small buckshot (either AAA or BB) is classified as short- to medium-range ammunition (5-15m), while large buckshot (SSG, SG, and LG) is effective at ranges between 15-40m, being able to stop a man up to 30m. When an SSG cartridge, which contains eight 9mm balls, is fired the result can be particularly devastating. With a spread of approximately 25mm per metre it has an effect comparable to a tight eight-round grouping fired from an SMG, but with only a single shot from a shotgun.

Currently the SAS use the shotgun in the CRW role to blow door hinges. The M203 40mm grenade launcher (underslung beneath the M16) can fire 'shotgun' anti-personnel rounds, and so has largely replaced the standard shotgun for use in the field. Civilian type pump-action shotguns have been largely discarded in preference to the more modern purpose-designed para-military models, such as the Remington 870P, Mossberg 500-ATP-8 or the Franchi special purpose automatic shotgun (SPAS) series.

The Remington model is in fact based on a design that was first introduced in 1949, and the 870P version has been officially adopted by the US Marine Corps as its combat shotgun, and the British Army for its Close Observation Platoons (COPs) in Northern Ireland. Both the Remington and the Mossberg models use aluminium in their construction to reduce weight, have pump-action mechanisms and folding-stocks which come over the top of the weapon. Both have been modified for the military from police/security models.

Holding the SLR as shown here gives the soldier the opportunity of firing from the hip or shoulder.

The Luigi Franchi SPAS models 11 and 12, on the other hand, were designed with just the military market in mind. They can, for example, operate in both pump-action and semi-automatic mode, giving the operator the option at the touch of a button, and the SPAS 12 can be fired with one hand if necessary. The SPAS 11 differs from the 12 in that the former has a rigid metal butt while the latter has a folding one. Both guns have short barrels, weigh 4.2kg, and fire all types of 12-bore ammunition from buckshot to very powerful armour-penetrating rounds.

They can operate in both pump-action and semi-automatic mode

Tear gas rounds can also be fired and there is an attachment to launch grenades to a maximum range of 150m. Using SSG rounds the SPAS can place 48 shots per second into a target at 40m and for extremely close quarter shooting the SPAS can be fitted with a shot diverter, which scatters the shot in a wider spread immediately on leaving the barrel. Another innovative design, this time from Britain, which increases the scope of the SPAS and other shotguns is the Hatton round. This ammunition was specially designed to take out door hinges, a useful asset to assault groups in hostage-release operations such as those conducted by the Regiment's CRW teams.

The modern shotgun can also be used against armoured vehicles. The SPAS has a 'penetrator' round, comprising a tungsten carbide discarding sabot which works on the same principle as some tank ammunition. This slug is capable of punching its way through light armour, such as that found on the sides of APCs. Weapons such as the SPAS, and the wide range of ammunition they can accommodate, allow the SAS greater scope in planning and carrying out both CRW and conventional warfare operations.

The SAS Regiment has used a wide variety of heavy support weapons over the years.

Left: Hand-held fury. The M16A2 assault rifle with an M203 grenade launcher attached.

Above: Firing a GPMG from the hip. The sling is used to support the weapon, which has a tendency to push back when firing.

Mortars, anti-tank and anti-aircraft weapons are all employed by SAS soldiers, providing them with an offensive and defensive capability. However, the employment of these systems rests on two factors. First, they must be the most effective systems available and second, they must be light hand-held models. SAS troops, and their individual patrols, must be able to carry their heavier weapons either on their persons, or mounted on Land Rovers or other small soft-skinned vehicles. The equipment must also be air-portable or capable of being dropped by parachute.

Deployed by air, boat, vehicle or inserted on foot, SAS patrols have a wide range of

Above: Members of 16 Troop, D Squadron, 22 SAS in Oman, 1958, with a variety of weapons.

weapons available to take with them: anti-tank weapons such as the Carl Gustav, LAW 80, M72 and Milan; 51mm and 81mm mortars; and Stinger and Javelin air-defence missile launchers — all systems which are capable of being man-packed.

Anti-tank warfare is not a primary role of SF soldiers such as the SAS, but the ability to knock-out tanks can be useful should units be tasked with a road ambush or similar mission. The Carl Gustav is a recoilless gun widely used by NATO and Scandinavian infantry units. Other lighter weapons, such as the M72 66mm Light Anti-armour Weapon (LAW), have been used extensively by SAS and airborne units in the past.

Above left: A GPMG in the sustained fire role. The link ammunition is being fed through the weapon by one of the crew.
Left: The GPMG mounted on a bipod. As the weapon is light enough to be carried for long periods, it can be used to provide SAS units on extended operations with long-range fire support.

Although the M72 system has been largely replaced by the newer 94mm LAW 80, the concept of these modern one-shot, throw-away weapons is more attractive than other heavier, re-usable anti-tank missile launchers such as the aged Carl Gustav, as the LAWs are light enough to be carried in addition to other weapons. The LAW 80 fires a fin-stabilised rocket with a high-explosive anti-tank (HEAT) warhead up to a range of 500m, and can be used for attacks on fortified targets such as bunkers, as well as armoured vehicles.

Medium-range anti-tank weapon systems such as Milan are also used by the SAS. Milan can be fitted with a MIRA thermal imaging device for night use, and is more accurate than the lighter systems. It also has a longer range —

1500m — but its weight of 23kg means that it can only be carried for specific missions.

The American FIM-92A 'Stinger' anti-aircraft missile system is another weapon that is man-portable and was used by the SAS offensively for the first time during 'Operation Corporate', the British campaign to re-take the Falkland Islands, although only one enemy aircraft was in fact shot down by this system. Fitted with an IFF (Identification Friend or Foe) system which allows the operator to positively identify his targets, it fires a heat-seeking missile to a maximum range of 8000m.

Mortars have always been used by the SAS as a means of providing indirect fire support. Two main types are currently in use with the British Army. The 51mm mortar can be carried and operated by one man and can drastically increase an SAS patrol's ability to engage targets indirectly, i.e. out of sight of the enemy.

The 51mm can fire a very effective HE bomb to a range of 750m. The larger 81mm, used by the SAS during their very successful raid on the Argentine air base on Pebble Island, West Falkland, is a much heavier and more powerful mortar. This weapon can fire an HE round to a maximum range of 5660m as fast as it can be loaded, although the normal rate is eight

The American man-portable Stinger surface-to-air missile (SAM) system. Used by the SAS in the Falklands and the Afghan rebels in Afghanistan, it is an extremely effective anti-aircraft weapon.

rounds per minute. However, the system is heavy and requires three men to carry it.

What weapons will the SAS be using in the twenty-first century? Undoubtedly they will be using the latest in weapons technology. This will mean lighter systems; making more use of plastics, composite materials and aluminium; weapons with greater accuracy and more reliable internal mechanisms; caseless ammunition and far more powerful propellants. All are presently being explored by arms manufacturers. It is conceivable that future SAS four-man patrols will be carrying Advanced Combat

Above: The Javelin SAM system in operation. The missile travels at a speed of over Mach 2.

Rifles such as the Heckler & Koch G11, and lightweight Glock handguns which contain modern composite materials. However, the emphasis within the SAS has always been very much on the men rather than the equipment they use. Careful selection and thorough training have been responsible for the Regiment's many successes throughout its relatively brief but active past, and this will no doubt remain the case in the future.

EXPLOSIVES AND SABOTAGE

David Stirling originally created the SAS to carry out sabotage missions behind enemy lines in north Africa. The Regiment's skill in the use of many different types of explosives today means that SAS four-man patrols can wreak large-scale destruction upon 'soft' targets — command and communications centres, depots and military equipment. As well as undermining an opponent's morale, these operations can tie down large numbers of troops in rear-area guard duties, thereby denying their use at the front.

Since the birth of the SAS in July 1941 the use of explosives and sabotage has formed an important part of the Regiment's operations. The SAS was originally conceived by David Stirling as a force that would launch small-scale raids behind enemy lines, causing the maximum amount of damage and disruption. The use of explosives constituted a large part of this strategy and once the unit became operational, in November 1941, Axis equipment, supplies and communications were regularly destroyed. Since that time all members of the SAS have received thorough instruction in the use of many types of

The moment of detonation. A vital bridge is destroyed by explosives laid by an SAS four-man patrol. Such work means patrols can cause damage out of all proportion to their size.

explosive materials.

SAS sabotage operations generally involve the disruption of enemy services, supplies or industrial output by violent means. Targets might be chosen because of their political, social or economic impact, or as part of an overall military strategy to confuse, hamper and tie down troops engaged in the static defence of vital installations or the search for saboteurs. Primary targets would include communications centres important to the enemy's command and control network which, because they are usually located far apart, require considerable military manpower to safeguard.

Primary targets would include communications centres

Other targets for sabotage attacks include military ships, vehicles and aircraft, bridges, roads, railway lines, telephone exchanges, power stations and supply depots. Targets are selected according to their strategic and tactical value. During a conventional conflict, for example, SAS patrols, liaising with their headquarters, may destroy a bridge that would otherwise have been used by the enemy to transport reinforcements to a sector of the front recently breached by friendly forces. Because of the importance of such a target, each member of an SAS patrol must be capable of selecting the correct type of explosive, computing the amount required, and knowing the exact position to lay the charges.

SAS soldiers are introduced to the art of 'cracks and bangs' during the Continuation phase of their training and, while a proportion of SAS troopers and noncommissioned officers (NCOs) go on to become demolition specialists, all undergo the basic introductory course.

An explosive is defined as a solid or liquid substance which, on application of a suitable stimulus (such as an exploding detonator), is converted in a very short time into other, more stable substances, usually gases, with the development of high pressure (an explosion).

This is the first lesson troops learn when introduced to the subject of military demolitions. Instruction in explosives is conducted at many levels but is generally less complicated than most soldiers believe. There are many types of explosive materials, just as there are many different jobs for them to perform. These explosives are known by a variety of chemical and military abbreviations: TNT (trinitrotoluene), Tetrytol, Composition C3 and C4 (plastic explosive/PE), Semtex, M118 demolition charge, Composition B, PETN, Amatol, RDX, Pentolite, Ednatol, and Dynamite.

At the Regiment's base on the outskirts of Hereford and in the surrounding military training areas, SAS students are instructed by their own experts and 'outside' specialists such as Sapper (Royal Engineer) NCOs. The Sappers, together with the ammunition technicians of the Royal Army Ordnance Corps (RAOC), are considered the best when it comes to the use of explosives.

There are two main types of explosive: low explosives and high explosives. In low explosives the process, once initiated, is one of combustion by burning, and is relatively slow. Examples of this type of explosive are gunpowder and cordite, the former used as an igniting agent for propellant charges — bullets, bombs and shells — while cordite is the propellant itself. The SAS are unlikely to use low explosives except when constructing improvised devices.

In order to set off high explosives an 'initiator' has to be employed

High explosives, the type used in the main by special forces (SF) for sabotage operations, function far quicker than low explosives, in a process known as detonation which results in a blast or shock wave. This shock wave has a great shattering effect and is especially suitable for weakening substances such as steel.

In order to set-off high explosives an 'initiator' has to be employed. There are in fact two main types of initiator: electrical and non-electrical. In a non-electrical firing system detonation is via a time fuze connected to a non-electric blasting cap — a thin aluminium tube 250mm long — which is embedded in the

Right: The aim when placing charges on a bridges is not to destroy the whole structure, but to cause its collapse by making it unstable.

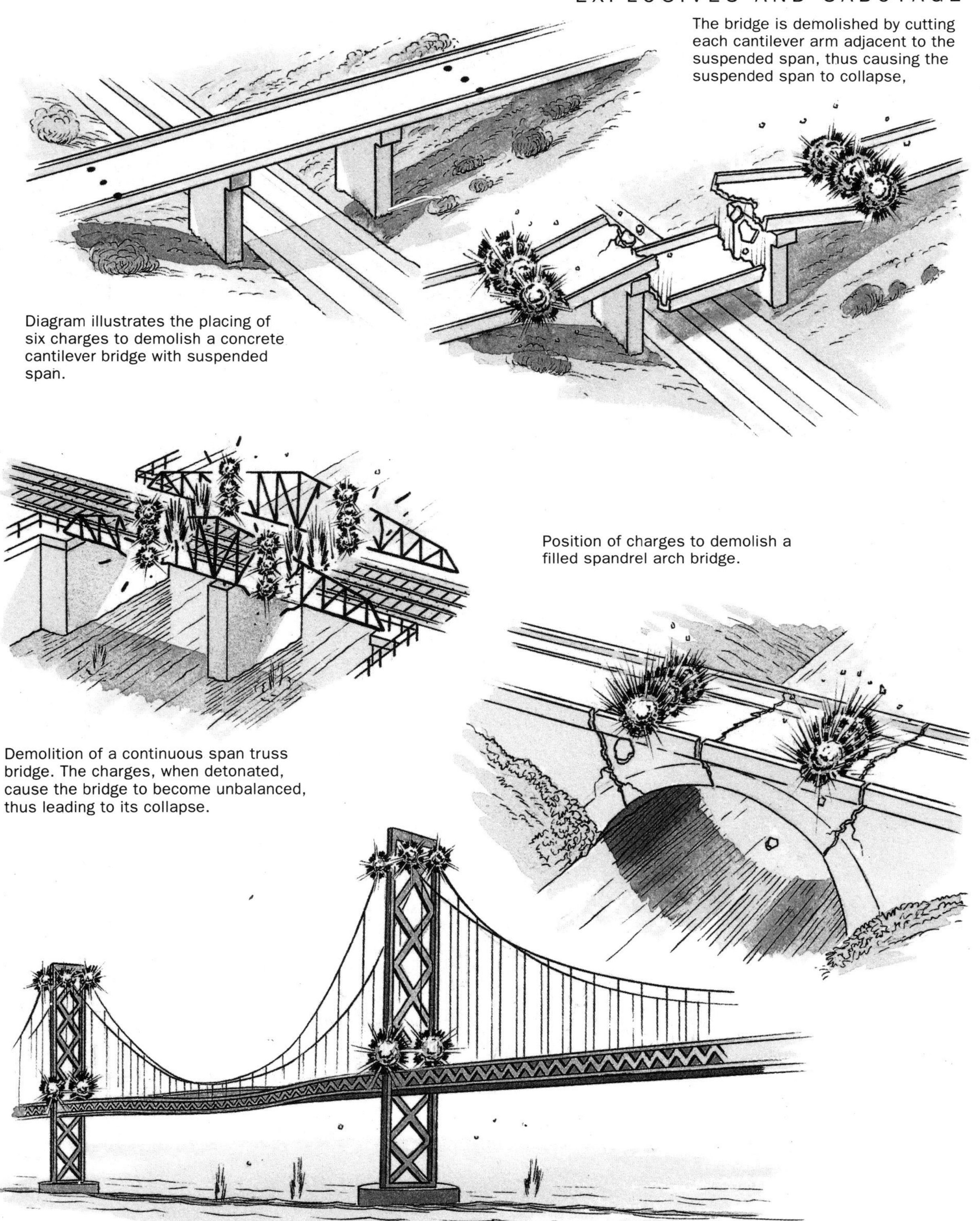

The bridge is demolished by cutting each cantilever arm adjacent to the suspended span, thus causing the suspended span to collapse,

Diagram illustrates the placing of six charges to demolish a concrete cantilever bridge with suspended span.

Position of charges to demolish a filled spandrel arch bridge.

Demolition of a continuous span truss bridge. The charges, when detonated, cause the bridge to become unbalanced, thus leading to its collapse.

To demolish a suspension bridge charges are placed on the towers slightly above the roadway, and on the cables near the top of the towers.

Above: A few kilos of well-placed explosive can bring a structure weighing thousands of tonnes crashing to the ground in a matter of seconds.

explosive charge. The blasting cap provides the detonating impulse required to ignite the charge. The fuze itself can either be a safety fuze or a detonating cord. Safety fuze is flexible cord about 5mm in diameter which consists of black powder encased in a fibre wrapping, itself covered with a waterproof material. Once lit, the fuze burns at a rate of 36-47 seconds per 300mm.

The fuze itself can either be a safety fuze or a detonating cord

The second non-electrical firing system is via detonating cord. 'Det cord' is reinforced prima-cord consisting of a small high explosive core protected by six layers of material, and detonates at a rate of 6-8000m per second. The det cord, like the safety fuze, is connected to a non-electric blasting cap. However, it cannot be detonated by a match or ignitor but must be fired by taping it together with two primers and a detonator which are then set-off either by safety fuze or an electrical current. The best analogy is the lighting a fire by traditional methods. The match is the fuze, the paper the detonator, the wood the primer and the coal the explosive charge.

In electric firing systems the blasting cap contains two wires which are connected in the cap by a bridge wire. When a current is passed through the wires the bridge wire becomes hot and ignites the charge within the cap, thus detonating the main explosive charge.

When placed, charges can be set off in three ways: by an SAS trooper lighting the fuze or detonating the charge in close proximity to the target; by means of a timer, whereby a charge detonates according to a pre-set timing device; and detonation by remote control. One example of an SAS timed device was the Lewes bomb, the invention of J.S. 'Jock' Lewes, an early member of the SAS. His bomb was a potent mixture of plastic explosive, thermite and lubricant which was detonated by means of fuzes called 'time-pencils'. In them, acid eroded through a metal wire in the pencil and released a spring which caused the explosion. Variations of wire thickness resulted in bombs with fuzes of half a minute to half an hour duration. They were ideally suited to destroying Axis aircraft and other equipment.

Currently SAS patrols use two main types of high explosive, either TNT slab explosive or C3/C4 plastic explosive, each with its own particular properties. Electric and non-electric firing systems can be used to initiate these explosives but the trick is to use the right system for the type of fuze, detonating cord or detonator, primer and explosive being laid. Once the target has been specified the type and amount of explosive is worked out, and planning for the positioning of the actual charges begins.

On actual operations explosives are often delivered to SAS teams by means of C-130 Hercules transport aircraft. Contrary to what many people believe, explosives are perfectly safe unless deliberately detonated by the correct method. For example, it is possible to hammer TNT into powdered crystals without it exploding, and it is quite normal for them to be delivered by parachute to troops operating in the field. Once the stores have arrived the SAS team tasked with the sabotage mission will begin the planning and preparation phase which will, if time and circumstances allow, involve some practice or dry runs on a similar 'friendly' target or a mock-up of the real thing.

Below: The ability to crater runways means SAS patrols can disrupt enemy air operations.

RENDERING AN ENEMY RUNWAY INOPERABLE

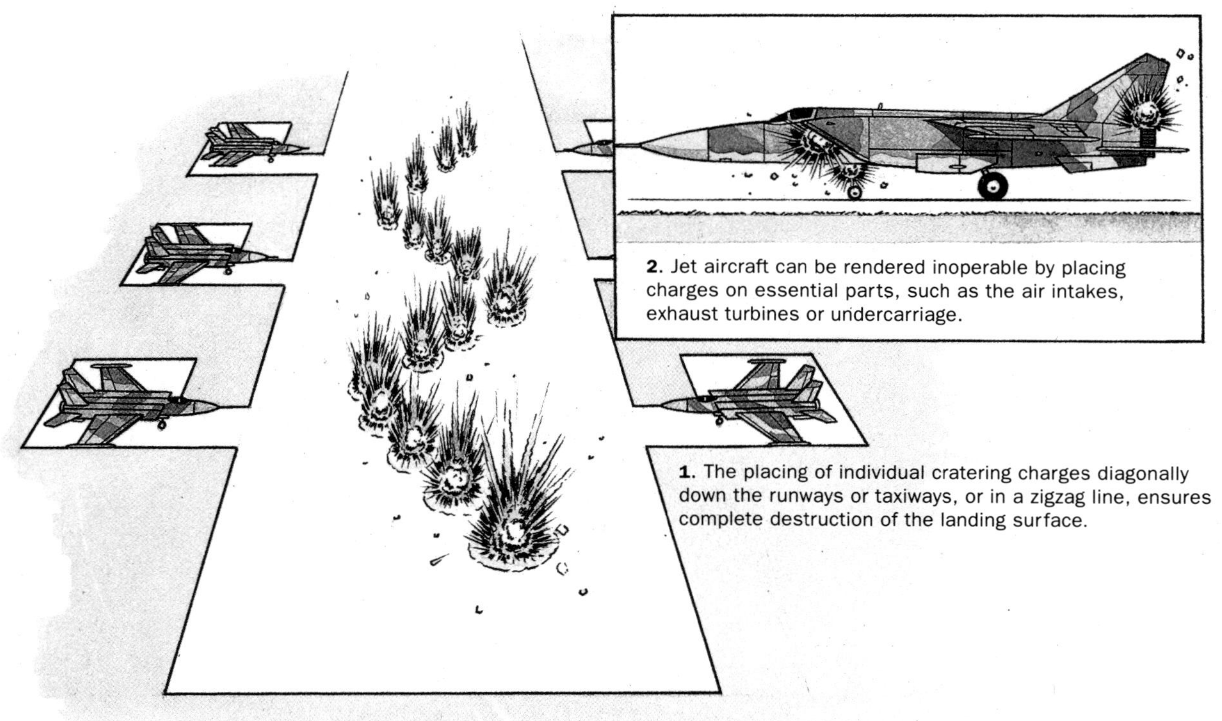

Rehearsals play an important part in the lead-up to any SF mission and the SAS set great store in being properly prepared. Equipment is checked and double-checked, contingency plans are discussed, final intelligence briefs are sought and scrutinised in detail. Little is left to chance. By the time the final go-ahead is given, the individual patrol members know exactly what they have to do in any event.

The SAS set great store in being properly prepared

Once the mission begins the patrol will make its way to the target undetected, avoiding contact with the enemy. When they are within striking distance of the target, part of the patrol will establish a lying-up position (LUP) while two soldiers, usually the patrol commander and the lead scout, will often conduct a close-target reconnaissance.

Exactly where to place charges for maximum effect depends on experience, the type of target being attacked, and the general tactical situation. Destroyed railway lines can be replaced relatively easily and it is therefore preferable to destroy bridges that support them or span them. Charges placed on architectural targets are designed to weaken the structure, causing them to collapse under their own weight. Vehicles and aircraft can be rendered inoperable by damaging their nose cones,

Argentine Pucara aircraft on Pebble Island. Note the damaged undercarriages and nose cones.

engines, axles, propellers or air intakes, and undercarriage. Ships can be sunk by blowing a hole in the hull just below the water line. These tasks do not require large amounts of explosives; what they do require is exact placement and an intimate knowledge of the destructive qualities of the explosive being used. During one very successful SAS raid in December 1941, for example, a four-man patrol destroyed no less than 37 Axis aircraft at Agedabia airfield in north Africa. They placed one small Lewes bomb, weighing no more than 400 grammes a piece, on each enemy aircraft.

Sabotage missions, like the one undertaken at Agedabia, are usually conducted under cover of darkness. While part of the patrol act as sentries the main party will move forward to lay the charges, which will have normally been constructed during the reconnaissance phase. If the target for demolition is a single structure such as a bridge, the patrol will place a number of charges in pre-planned positions, and arrange for them to explode simultaneously. Targets such as bridges are best blown manually if the tactical situation allows, and at a time when they are in use by military traffic. Although this maximises the effect of the attack, it can also draw attention to the patrol carrying out the attack, thus increasing the risk of death or capture.

Sabotage missions are usually conducted under cover of darkness

In certain circumstances it may be decided to openly attack the target once the element of surprise has been lost. For other missions timers are used to delay the charges from exploding until the patrol laying them has had a chance to withdraw from the area. An example of the former type of operation occurred when an SAS squadron attacked the Argentine airfield on Pebble Island during the Falklands campaign. This operation involved an approach under cover of darkness and, once the enemy realised the airfield was being attacked, the use of integral and support fire to suppress the opposition while the targets were being engaged. Aircraft were destroyed by a combination of explosive charges and rounds from anti-tank weapons, with the SAS conducting a withdrawal under covering fire supplied by their own mortars and naval gunfire support.

The use of delayed explosive charges, or demolitions involving the use of a trigger mechanism such as a pressure plate, are all tried and tested techniques that have been employed by members of the SAS Regiment over the years. With new types of explosives being constantly developed, the SAS will have the means to wreak ever-greater havoc on numerous targets situated behind enemy lines, thus increasing their potential as a war-winning unit.

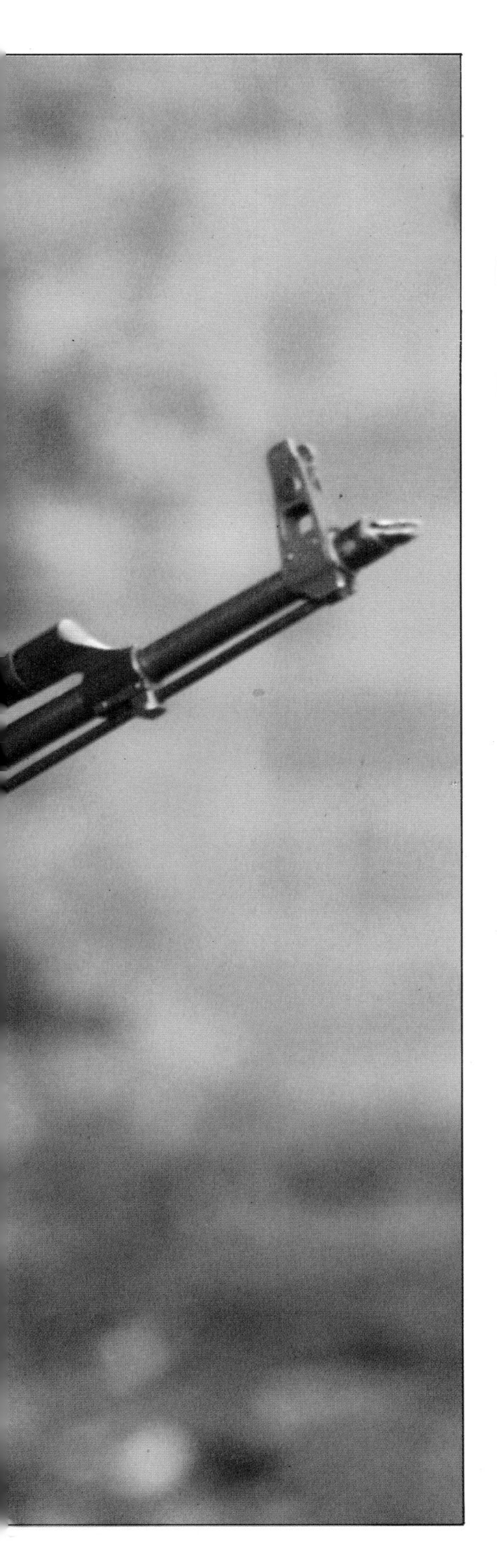

COUNTER-TERRORIST TACTICS

To combat the growth of international terrorism and to devise effective counter-terrorist tactics, the SAS established the Counter Revolutionary Warfare Wing at Hereford. The skills and techniques taught there ensure that the Special Air Service has one of the world's most effective counter-terrorist units. In addition, the Troubles in Northern Ireland have added an extra dimension to the Regiment's counter-terrorist operations, with teams frequently being deployed to the area to fight the IRA.

Together with sabotage terrorism is one of the primary weapons of the revolutionary. There are dozens of terrorist groups throughout the world, with many being highly trained and well equipped. They range from nationalist, such as the Irish Republican Army (IRA) and the Basque separatists ETA, to extremist organisations, such as Germany's Red Army Faction and Italy's Red Brigades. However, they have one thing in common: they are committed to the use of violence in the furtherance of their aims. Guerrilla units or special assassination squads can attack both specific individuals and innocent groups using bomb or bullet, take hostages, hijack aircraft,

An IRA gunman with a Kalashnikov AKMS assault rifle. Though small — it has no more than 50 full-time 'soldiers' — the IRA is an extremely effective terrorist organisation.

Above: Members of 'Pagoda' Troop, armed with MP5 submachine guns, on the roof of the Iranian Embassy, just prior to their assault.

and commit acts of arson and extortion.

The 1960s and 1970s witnessed the growth of the international terrorist network, with criminal acts being committed in the name of various causes. The Palestinian issue was the cause of numerous hijackings and murders, such as the take over of an Air France airliner in 1976 by Palestinian and Baader-Meinhof terrorists, which was to result in the spectacular Israeli operation at Entebbe. The 1980s witnessed several IRA attempts to assassinate the British prime minister; attacks on Western targets by Shi'ite Moslem groups in Lebanon; and various kidnappings in Western Europe by extremist groups, such as the abduction of General Dozier by the Italian Red Brigades.

There are currently some 800 terrorist organisations throughout the world

There are currently some 800 terrorist organisations throughout the world, and many of them are interconnected. In addition, several countries provide havens, financial support and training facilities for international terrorist groups. Libya, for example, is estimated at providing financial backing for over 40 terrorist organisations. Many Palestinian terrorist groups receive aid from the Soviet Union, China, North Korea, Vietnam, Iran and Cuba, as well as several Arab states — Algeria, Libya, Egypt, Yemen, Saudi Arabia, Iraq, Syria and Jordan.

In an effort to defeat international terrorism many Western governments have formed counter-terrorist (CT) units, for example France's *Groupement d'Intervention de la Gendarmerie Nationale* (GIGN); Spain's Grupo *Especial de Operaciones* (GEO); and Italy's *Nucleo Operativo Centrale di Sicurezza* (NOCS). Many of these groups liaise with each other concerning tactics, training and developing equipment for CT operations. The SAS, for example, has often provided expertise and specialist equipment for hostage-rescue missions to its European counterparts. In addition, the various intelligence agencies, such as the Central Intelligence Agency (CIA) of America and Britain's MI5 and MI6, often share information concerning the various activities of international terrorist groups.

The success of 'Pagoda' Troop, the codename for the team involved in ending the Iranian Embassy siege in 1980, was testimony to the effectiveness of the Regiment's hostage-rescue training and tactics. Although the Counter Revolutionary Warfare (CRW) Wing at Hereford was officially created in 1973 when the Regiment formally adopted the CT role in response to the killing of 11 Israeli athletes at the 1972 Munich Olympic Games by the Palestinian group 'Black September', the SAS had in fact been exploring this area of operations before it had first been sent to Northern Ireland in 1969.

The previous year international terrorism had taken a turn for the worse with a series of aircraft hijacks by the Popular Front for the Liberation of Palestine (PFLP). The need for a highly trained and capable CT unit to deal with such situations was realised by SAS commanders, and steps were taken to prepare the Regiment for the inevitable fight against international terrorism. The SAS would have three specific roles in combating terrorism: intelligence gathering and collation; responding directly to specific terrorist threats; and conducting pre-emptive strikes on terrorist cells. As a result the CRW Wing at Bradbury Lines, the old SAS camp on the outskirts of Hereford which was later replaced by the specially-built Stirling Lines, was established to explore the tactics, training and technology needed to defeat the terrorists.

Below: Four views of an SAS soldier's Counter Revolutionary Warfare kit. The type of Heckler & Koch carried varies according to the mission.

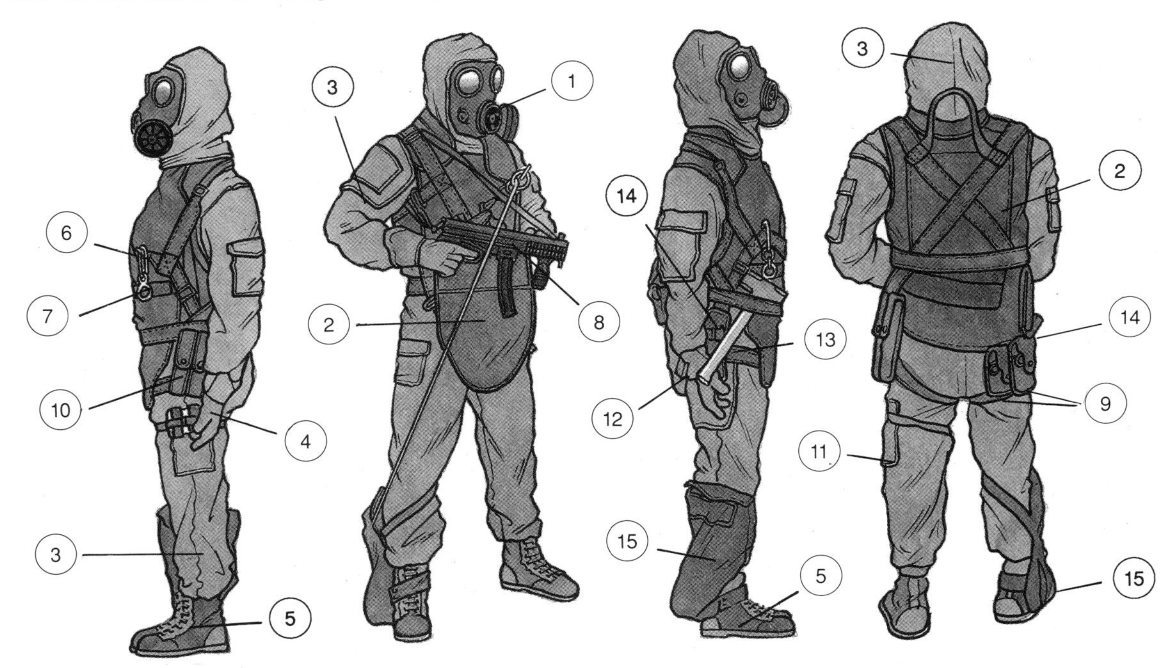

1. S10 RESPIRATOR
2. BODY ARMOUR
3. FLAME-RETARDANT SUIT
4. FLAME-RETARDANT GLOVES
5. ANKLE-HIGH, RUBBER-SOLED BOOTS
6. 'D'-SHAPED RING
7. FIGURE-OF-EIGHT DESCENDER
8. MP5K
9. STUN AND CS GRENADES
10. POUCH FOR THREE MP5K MAGAZINES
11. TWO 13-ROUND MAGAZINES FOR HIGH POWER STRAPPED TO LEFT THIGH
12. ONE HIGH POWER MAGAZINE STRAPPED TO RIGHT WRIST
13. SLEDGEHAMMER TO BREAK DOWN DOORS, WINDOWS, ETC.
14. HIGH POWER HANDGUN IN HOLSTER
15. BAG TO HOLD ABSEILING ROPE SO AS NOT TO ALERT ANYONE BELOW POINT OF DESCENT

Above: A two-man team neutralises armed terrorists during a hostage-rescue operation.

The primary aim of the Wing has always been to devise ways of rescuing hostages alive, rather than the elimination of their captors. However, in most cases the only way to secure the safety of hostages is to ensure that the terrorists are killed outright. During an operation the individual SAS trooper is taught to take no chances. Given a clear shot at a terrorist he will often empty an entire magazine into the target. There is a good reason for this: the modern terrorist uses remote detonation devices that require only the push of a button to detonate a bomb. A double tap (two shots fired in quick succession) on its own is often insufficient to stop a determined terrorist from detonating hidden explosives, or pulling a weapon which can then be used to kill hostages or CT personnel. Sustained and accurate firepower keeps the terrorist's hands away from his body, thus making it impossible for him or her to reach for a gun or push a button.

The SAS employ body-line shooting in their CT operations. Where possible shots are always aimed at the trunk of the body as it contains the vital organs — the heart, lungs and liver. Head shots usually require only one bullet to kill the terrorist outright, but the target area is much smaller. In addition, in a darkened smoke-filled room full of hostages and terrorists, the chances of shooting accurately are greatly diminished. Nevertheless, SAS troopers are trained to use head shots if that is the only target presented to them, as when a terrorist is taking refuge behind a hostage or is hiding behind cover.

As the terrorist threat increased during the 1970s the CRW Wing was expanded, more instructors were trained, and all SAS 'Sabre' Squadrons received CT training on their return from operational tours abroad. Today,

Right: A CRW assault group member engaging targets with a Remington 870P pump-action shotgun. These weapons have been used by the SAS during CT operations in Northern Ireland.

at any one time, a squadron is on 24-hour standby for CT operations. It is divided into four operational troops, known as Special Projects Teams, of one officer and 15 other ranks. These are further broken down into four-man operating teams and, during assaults, these four-man units work as two-man teams.

The first CQB range had in fact been established by A Squadron in Aden

As the Regiment's expertise in CT techniques increased, training became tougher and muc more technical. The emphasis was on Close Quarter Battle (CQB) skills and a special building was erected at Hereford to test the SAS troopers' technical and practical knowledge to the limit. The first rudimentary CQB range had in fact been established by A Squadron in Aden during the SAS campaign in that country (1964-67). The building at Hereford, known ominously as the 'Killing House', comprised a sophisticated series of indoor ranges designed to mirror any conceivable interior which members of a Special Projects Team might encounter during an actual hostage-rescue mission, from a semi-detached, two-storey building to multi-roomed government complexes; from railway carriages to aircraft cabins and cockpits.

Remotely-operated, spring-loaded figure targets representing both hostages and captors

The front of the Iranian Embassy, 5 May 1980. Note snipers and the two hooded SAS soldiers.

can be arranged around the rooms and set to pop-up to be either ignored or engaged, as the case may be. During the six-week period allocated to CRW training, SAS troopers can be expected to fire literally thousands of rounds on these ranges, in the process developing extraordinarily rapid reflexes and an uncanny ability to almost instantly appraise the immediate tactical situation.

Most hostage-rescue exercises take place in the 'Killing House' using live ammunition and pyrotechnic devices, although the SAS is not restricted to Hereford for its CRW training venues. Over recent years the Regiment has acquired a number of additional training aids including railway carriages, an airliner and wheeled passenger vehicles. Neither are the SAS CRW experts confined to the UK when it comes to practising CT techniques. Close links have been established with other specialist CRW units such as Germany's para-military *Grenzschutzgruppe 9* (GSG 9) and America's Federal Bureau of Investigation (FBI) with its Hostage Response Team. Intelligence information on international terrorist cells, their techniques and composition, is collected, collated and disseminated throughout the CRW fraternity. In addition, since the Iranian Embassy siege in 1980, the SAS have compiled a comprehensive computer database to aid them in their CT operations, concerning such things as the thickness of doors in buildings that are potential terrorist targets, and the design of all aircraft that are currently in service.

Most exercises take place in the 'Killing House' using live ammunition

SAS CRW training is considered amongst the best available anywhere in the world. It is as realistic as possible in keeping with the Regiment's maxim 'Train Hard - Fight Easy'. It is also dangerous. Until relatively recently live-fire exercises in the 'Killing House' involved the use of live 'hostages' played by SAS soldiers interspersed with terrorist figure targets or mock-up dummies. The CRW assault team would enter the room and immediately have to identify the terrorists from the 'good guys', taking-out the former while leaving the latter unharmed.

This system worked very effectively, especially with regard to inducing stress among the members of the squad conducting the assault, until an SAS noncommisioned officer (NCO), playing a hostage, moved inadvertently and was shot dead by one of the assault team. Following an inquiry into this tragic accident it was decided to make increased use of technological aids, and the 'Killing House' now comprises a hostage-rescue area consisting of two separate rooms. One, containing the 'terrorists' and their 'captives', is connected by a sophisticated camera system to a second room which is attacked by the assault team. All events taking place in one room are simultaneously projected onto a life-size wraparound screen in the other and vice-versa. This allows live 'terrorists' and 'hostages' to be used, seen and engaged without actual risk by the assault team. Both sets of images are projected onto bullet-absorbent walls which allow both groups to fire at each other in a realistic three-dimensional scenario.

Each two-man team has its own rooms or areas to deal with

The aim of the CRW Wing is to finely hone the individual SAS soldier's weapon and firing skills including shooting on the move; rapid magazine changes and malfunction clearance procedures; rapid target identification and acquisition; and accurate shot placement impacting on the body-line. In a hostage-rescue situation each two-man team has its own rooms or areas to deal with, and each trooper has his own arc of fire. He must be adept at calculating his arc of fire to encompass any area in which he is operating at a particular moment. For example, in a corridor the field of vision is limited and consequently the arcs of fire are reduced, but carry on down the corridor and you have doors, each one a possible threat, that must be approached and cleared, under cover of your comrade. Once inside a room the arcs spread out to cover 360 degrees for all-round defence, as there is always the threat of being engaged from behind.

One method of countering the natural tendency to only engage targets to the front and side was developed by the Regiment's CRW instructors in the early days of the 'Killing House'. This involved a machine that fired tennis balls at head height from a concealed position opposite the targets at high speed. Failure to scan the full 360 degrees after the first targets were engaged would result in a thump in the back of the neck.

In addition to learning individual skills, at Hereford troopers will receive instruction in the tactics of entering and clearing buildings, aircraft cabins, railway carriages and other

Above: All SAS counter-terrorist personnel are experts in abseiling skills and weapons handling. The ability of team members to enter buildings via windows as well as doors gives commanders greater tactical flexibility during hostage-rescue operations.

structures. The first phase consists of isolating the building containing the terrorists and their hostages, and securing the perimeter. For example, all existing telephone services will be cut off to eliminate the terrorists' access to the media and public officials. They will be able to communicate only with the official negotiator.

The SAS unit will divide into two groups, an assault group to enter the building, and a perimeter containment group consisting of marksmen who provide a cordon around the target. At the Iranian Embassy siege, when six terrorists of the Democratic Revolutionary Movement for the Liberation of Arabistan took 26 innocent people hostage, the police also deployed snipers to cover the building. In such a situation the marksmen fulfil several important functions: maintaining overall surveillance; neutralising any persons who pose a threat to a hostage or assault team member; and providing cover and assistance for the assault group when the attack is launched.

As negotiations commence to resolve the situation peacefully, the SAS will prepare a plan of assault. The collection of good intelligence at this stage is crucial to the ultimate success of the SAS assault plan, should it be used. At Princes Gate electronic intelligence specialists from C7 (the police technical support branch) installed surveillance devices and microphones through the walls of the buildings adjoining the Embassy, allowing the SAS to build up an accurate picture of their target while rehearsing for an assault. They spent five days planning the operation. A scale model of the large 50-room Embassy was constructed, with every detail of its layout being provided by the building's caretaker. In addition, one of the hostages, suffering from stomach pains, had been released by the terrorists. He provided much valuable information concerning the whereabouts of the hostages and their captors.

Below: A GSG 9 soldier armed with an MP5SD, a silenced version of the MP5. Note the telescopic sight and the 15-round magazine.

Members of the Italian counter-terrorist unit *Nucleo Operativo Centrale di Sicurezza* (NOCS), shown here, are experts in Judo and Karate.

When negotiations break down then an assault becomes necessary. The success of the operation ultimately depends on speed and coordination, with all assault teams entering the building simultaneously. Distraction techniques are employed to disorientate the terrorists and give the CT unit an advantage. In 1977, for example, seven South Moluccan terrorists held 49 people hostage on a Dutch express train in Holland. The siege was eventually resolved when Dutch Marines successfully stormed the train. Just prior to the assault six F-104 Starfighter jet aircraft flew low over the railway carriages, kicking in their afterburners to distract the terrorists and keep the hostages' heads down.

At Princes Gate the SAS plan was to blow in the first-floor armoured plate window at the front of the building using frame charges, with a four-man team crossing from an adjoining balcony to enter the Embassy. Two four-man teams would also rapidly abseil down from the roof at the rear to the first- and second-floor windows, and also blow open the back door. The front window would be blown in first as a signal to the abseil teams to begin their assault, and as a means of distracting the terrorists towards the general direction of the blast. Frame charges, containing explosive charges in strip form, are especially useful for hostage-rescue situations requiring a method of rapid entry. They can be used to blow precision holes through steel, brickwork, armoured glass, windows and doors.

The pointman leads the entry team during the approach

The four-man assault team consists of a team leader, a pointman, a defenceman and a man responsible for rear security. The leader is in charge of his team's assault, and he also liaises with the perimeter containment group. The pointman leads the entry team during the approach and lays the frame charges; the defenceman provides security for the pointman; and the last man always serves as second-in-command and carries any additional equipment, such as a Remington pump-action shotgun.

Even the most well-rehearsed plans can go wrong, and so the SAS always allow for the unexpected. At Princes Gate, as the troopers descended on their ropes, a window was inadvertently smashed, and one man became entangled in his rope opposite a second-floor window. Unable to use their explosives to effect an entry because of the risk to this man, the SAS smashed their way in using their sledgehammers, which are always carried for such an eventuality. At the front all went according to plan; the frame charge had shattered the window, allowing the SAS team to enter.

The assault teams themselves will also be carrying CS gas

Once inside the building, the assault teams will quickly and systematically begin to clear the rooms, hallways and stairs of enemy personnel. Snipers from the perimeter containment group will provide covering fire and, as at Prince Gate, will shoot CS gas canisters through the windows. The assault teams themselves will also be carrying CS gas. In cloud form CS gas causes immediate debilitating effects: a severe burning sensation in the eyes, tears, coughing, difficulty in breathing and a tightness in the chest. The eyes close involuntarily, the nose runs and moist skin burns. As a consequence, those affected are incapable of effective concerted action.

Room-clearance drill is quick and simple: if any rooms have their doors locked or barricaded then one member of the four-man team will blow off the door hinges using a shotgun; the door will be kicked in and stun grenades tossed inside the room; a two-man team will immediately follow, killing any hostiles with double taps and three-round bursts — the action is over in less than four seconds. The team will then immediately move on to other rooms. Each team will have its own area of the building to clear, assigned during the planning phase. For assaults on railway carriages and aircraft cabins, all entrances are attacked at the same time to achieve multiple simultaneous entry.

The stun grenade, or 'flash-bang', is one of the most useful items in the SAS CT armoury and was devised by the CRW Wing at Hereford. The CRW Wing, together with the Regiment's Operational Research Wing, is responsible for the development of equipment and weaponry for use by the Regiment in its CT operations. They have developed such things as special assault ladders, vans with hinge-mounted ladders for rapid second-storey entry, numerous night optics, thermal imagers and endoscopes (a long slender medical instrument used for examining the interior of hollow organs, but also useful for CT operations).

The stun grenade is a non-lethal device ideally suited to hostage-rescue missions. It is a mere 15cm high and 10cm around, weighing just over 250g. It contains a cocktail of magnesium powder and fulminate of mercury. When the pull ring is removed and the device thrown the mercury fulminate, a percussion explosive, detonates. This results in an extremely loud bang which also ignites the magnesium, generating intense light (a 50,000 candlepower flash). The two reactions coupled together produce the 'flash-bang' effect, the high candle and decibel levels cause extreme disorientation among the unprotected for up to 45 seconds, time enough for the assault team to neutralise the terrorists and secure the safety of the hostages.

The operation was a success with three of the terrorists shot dead

SAS 'flash-bangs' were first used operationally during the storming of a Lufthansa Boeing 737 by Germany's GSG 9 in October 1977. The airliner had been hijacked by members of the Red Army Faction terrorist organisation, and had eventually ended up at Mogadishu airport in Somalia. The SAS Regiment's CRW specialists had conducted a number of training exercises with GSG 9 and the latter now requested their assistance. The SAS supplied two men, Captain Alastair Morrison and Sergeant Barry Davies, who joined the 30-man

Left: SAS CRW teams carry sledgehammers to aid them in entering buildings.

GSG 9 team in Somalia, taking with them a quantity of stun grenades. After the terrorists ruthlessly murdered the aircraft's captain, one of the 86 passengers and 6 crew being held, it was decided to storm the aircraft.

After using a fire on the runway to lure the terrorists onto the flight-deck, thus reducing the direct threat to the passengers, rubber-coated ladders were stationed next to the emergency exits over the wings and against the doors to the fore and aft of the Boeing. All doors were blown simultaneously with explosives charges, and stun grenades were thrown inside the aircraft by the two SAS men just prior to it being assaulted by members of GSG 9. The operation was a stunning success, with three of the terrorists shot dead and the fourth very seriously injured. It was GSG 9's first victory and was helped to a large extent by the 'flash-bangs' brought from SAS headquarters at Hereford.

The initial target of any assault will be the room where the hostages are being held. There are two reasons for this: the terrorists may begin to kill their prisoners, and the building will quickly fill with CS gas and smoke as a result of any fires caused by the explosions. At Princes Gate the majority of the hostages were held in the Embassy's telex room. One hostage had already been killed and two more wounded by the time the first team burst into the room.

A spare magazine is often worn on the left wrist for a rapid magazine change

Entry teams require proper clothing and equipment for the successful conclusion of CT operations. SAS troopers wear a black one-piece suit, manufactured from fire-retardant material such as Nomax, together with black combat boots. Over the suit a bullet-proof sleeveless black jacket is normally worn, which comprises a casing into which ceramic Kevlar armoured plates are inserted front and rear. Similar to flak-jackets in basic design, these items are tailored to meet an individual's particular requirements, and include pockets to contain ancillary equipment such as radios and grenades. Gloves are worn to ensure a good grip on weapons and other items of equipment. In addition, a belt rig, usually made of ballistic nylon or tough hide, is worn around the waist and comprises a personal selection of pouches. A drop holster rides low on the right leg, with spare magazine pouches worn on the other leg. In addition, a spare magazine is often worn on the left wrist for a rapid magazine change.

As at Princes Gate, abseiling is a method of entry frequently employed by CRW personnel, and so integral or strap-on seat-type harnesses are often worn. Used in conjunction with stabilising descenders, quick-release fittings and a rope-stowage leg bag, these harnesses are similar in design to those worn by civilian rock-climbers. Other items of equipment worn by individuals include spare magazines for SMGs and handguns, bolt cutters, hand hammers, single-headed axes and glass cutters.

Above: SAS troopers at the Iranian Embassy.
Right: Cut-away of a stun grenade.

Experience has taught that only materials of the highest quality are suitable for hostage-rescue operations. The team member who got trapped on his abseil rope at the rear of the Iranian Embassy did so because 'Pagoda' Troop had insufficient rope, and therefore had to purchase an additional amount. The new rope was of an inferior quality and overheated during the assault due to friction caused by the weights imposed upon it; it then ravelled into a knot and trapped the trooper. Other items of equipment worn by individuals include spare magazines for SMGs and handguns, bolt cutters, hand hammers, single axes and glass cutters.

Sewn-on patches for scabbards complete the rigs and are attached either to the upper-arm or thigh, or the upper-chest of the body armour waistcoat. Knives are employed to cut troopers free from snared abseil ropes rather than as anti-personnel weapons, although they

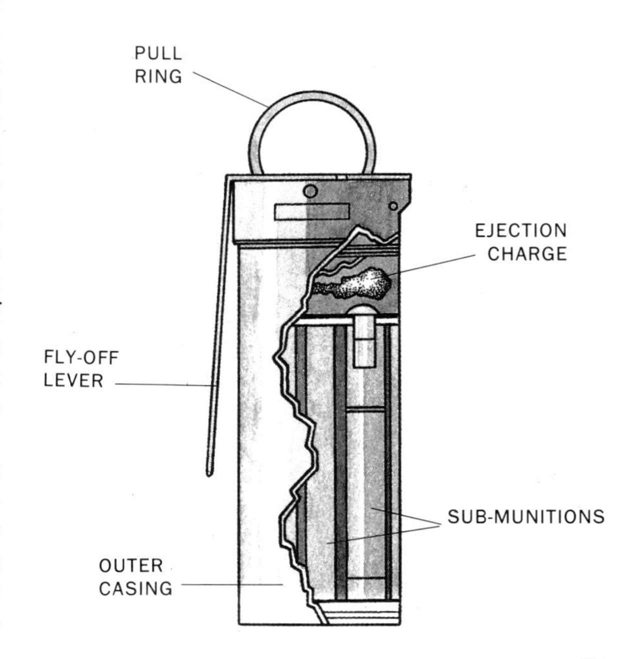

MOGADISHU AIRPORT, SOMALIA

18 OCTOBER 1977 00.06 HOURS

could no doubt be used in the latter capacity as a last resort.

SAS weapons must work first time and every time

Respirators, the correct term for 'gas-masks', are worn for many types of CRW mission, and ensure the assault teams are not affected by CS gas or smoke. The S10 version has replaced the S6 model used in the Iranian Embassy operation, the major difference being the eyepieces, which are now round and flat as opposed to oblong and shaped. Darkened filters are attached to protect the wearer's vision from the severe effects of stun grenades. In addition, the respirators also contain internal microphones which, together with either headsets or earpieces, provide the usual communications

Above: Recreation of GSG 9's assault upon the Boeing 737 aircraft at Mogadishu airport.

requirements. This is vital for CT operations as it means that leaders of the entry teams can remain in contact with their men throughout the assault. Commanders outside the building can also be given an accurate assessment of unfolding events, thus retaining effective overall command and control of the operation. During the preparations for the assault, and during its execution, hand and arm signals will be used by SAS teams. They eliminate unnecessary talking and enhance overall noise discipline.

A flash-hood, worn over the respirator, provides both protection from fire and blast, and helps secure the straps of the mask and communications devices. The black mask and hood, as well as being functional, add to the general fearsomeness of the assault team, thus giving it a split-second edge.

For hostage-rescue operations individual troopers will usually carry a submachine gun (SMG) as a primary assault weapon, with a handgun as a backup. The weapons carried by SAS assault teams must fulfil certain criteria: they must work right the first time and every time; they must be as small as possible to wield in confined spaces; and they must have good stopping power. The Heckler & Koch range of SMGs and the Browning High Power handgun fulfil all of these requirements.

The MP5 series provides the SAS with a comprehensive range of weapons for CT work. For example, the MP5SD is a suppressed version of the MP5 which gives entry teams a weapon with a low noise level and reduced muzzle flash, making it difficult for terrorists in a darkened room to accurately return fire. It also mounts a very powerful laser lock system (which have replaced the mounting of small electric torches on SMGs for use in darkened rooms), giving the SAS a sighting system which can constantly show the firer where the burst is going to strike, without him having to shoulder the weapon. It is an extremely good combination for close quarter battle.

Troopers will usually carry a submachine gun as a primary weapon

The choice of the correct type of ammunition is also crucial to the success of a CT operation. Special ammunition and pyrotechnic devices have therefore been developed, for example ammunition which reduces the risk of fire when rescuing hostages from confined inflammable areas like aircraft cabins. The SAS use 9mm calibre weapons (MP5s and High Powers) for close quarter combat during hostage-rescue missions. The 9mm bullet in itself is an excellent round. It is smooth, with

Below: Debussing from a helicopter, members of GSG 9 quickly establish a roadblock. Such rapid drills are practised by all counter-terrorist units.

no sharp angles, small and has excellent stopping power. Its one drawback with regard to hostage-rescue work is that it tends to shoot straight through the target and keep on going. This means that any hostages situated behind a terrorist being shot would also be hit. When the SAS shot three IRA terrorists at Gibraltar in March 1988 they used standard 9mm ammunition. One of the three, Mairead Farrell, was hit by three bullets in the back. They all passed through her chest and exited at the front.

For hostage-rescue work the SAS use special 9mm bullets, designed to kill the terrorist but remain in his/her body, thus minimising the risk to any hostages. These so-called 'fourth generation rounds' come in two main categories. First, the new frangible round is designed to break-up in the human body. One particular model used is the 'Splat' metal-and-polymer mix bullet, which can penetrate soft-to-medium cover without breaking up until it exits the far side, and so still strikes the target. Second, there is the powerful Accelerated Energy Transfer (AET) round. These bullets are shaped to have an effect when they hit the body similar to that of a swimmer performing a belly flop dive into water. One such round, the THV, with a concave-shaped tip, has low ricochet, low recoil, good stopping power and armour piercing qualities. It causes massive tissue damage and is considered a superb specialist bullet.

Such specialist rounds and the general paraphernalia of hostage-rescue missions are rarely used in Northern Ireland where, since 1969, the SAS have been involved in another type of anti-terrorist activity: the war against the IRA. It is a vicious, never-ending war, but

Below: In Northern Ireland the SAS live in cramped, uncomfortable conditions, very similar to the quarters shown here.

Above: Unlike this Army patrol, the SAS keep a low profile when operating in Northern Ireland.

one to which the Regiment is in many ways ideally suited. The tactics used in Ireland are the result of lessons learned from earlier counter-insurgency (COIN) operations in the jungles of the Far East during the 1950s and 1960s. These earlier operations involved the extensive use of highly-developed individual, patrol and troop skills, with missions undertaken behind enemy lines as well as classic search-and-destroy operations and ambushes.

The SAS campaign in Aden also yielded valuable lessons which prepared the Regiment for its work in Northern Ireland. In the city of Aden itself, in the so-called 'Keenie-Meenie' operations, SAS soldiers were disguised as Arab civilians. Teams would consist of two or three men, who would sometimes take with them a comrade dressed in British Army uniform, or European dress, to act as bait to lure the enemy terrorists (specifically Yemeni-trained assassins) into an ambush.

COIN operations have been modified by the SAS for use in a 'home' environment. There are no search-and-destroy missions in Northern Ireland, nor are there authorised cross-border operations, at least not officially. But some traditional SAS skills, such as surveillance, intelligence gathering and setting-up ambushes, have stood the Regiment in good stead in their war against the IRA.

The ambush is the favoured SAS tactic in Northern Ireland

The key to successful CT warfare in Northern Ireland is the continuous collection of intelligence and its timely dissemination between the various agencies of the security forces. In Ireland the SAS is supported by the Royal Ulster Constabulary (RUC) and the various Army agencies, such as the 14th Intelligence Unit. Indeed, the SAS has trained many of 14th Intelligence Unit's recruits in the art of covert surveillance, communications and agent running. Similarly, accurate intelligence analysis, together with information gathering, are finely-honed SAS tactics and can often lead to a particular terrorist target being identified and an ambush being laid. The SAS observe suspects, isolated farms and houses often for weeks at a time. Each man is equipped with a 'bingo book' containing lists of wanted men, missing vehicles and suspect addresses. In South Armagh, for example, the SAS operate in plain clothes, spending days at a time in specially concealed dugouts near cross-border routes.

Good intelligence, close cooperation between the various agencies, and patience can result in spectacular success. The ambush is the favoured SAS tactic in Northern Ireland, and it is one which the Regiment has become adept at carrying out. In April 1987, for example, the RUC were alerted to the theft of a JCB mechanical digger in East Tyrone. This in itself was not unusual. However, a year earlier an RUC station at The Birches had been attacked by the IRA, who had used a stolen JCB digger, with a large bomb in its bucket, to ram the station's gates. Surveillance of known IRA members thought responsible for similar

Above: Recreation of one of the most successful counter-terrorist operations undertaken by the SAS in Northern Ireland: the ambush at Loughall.

operations was intensified. When the JCB was located at a remote, disused farm some 16km from the RUC station at Loughall, it seemed as if the security forces had identified the vehicle and the target of a future terrorist attack. Surveillance was established and RUC officers witnessed the transportation of explosives to the farm. The security forces now had a good idea of the type of attack and the probable target, but no date. Fortunately, however, the RUC intercepted a phonecall made by an IRA man concerning the timing of the attack.

A classic box-type ambush would be sprung on the terrorists

The local SAS commander, liaising with his RUC counterpart, moved some of his men into the now deserted police station itself, while others deployed in the surrounding area. They established concealed positions both within the station compound itself and in the hedgerow opposite, with further cut-off positions down the road either side of the isolated police outpost. RUC snipers took up position to cover the SAS soldiers. A classic box-type ambush would be sprung on the terrorists when they arrived. The SAS and RUC both settled down patiently to wait.

On 8 May the terrorists, under surveillance, were seen to be preparing to move. Shortly after 1900 hours that evening, an eight-man IRA active service unit (ASU) approached the area to the front of Loughall police station in a stolen van and the JCB digger with the bomb in its bucket. The five men in the van then proceeded to lay down a barrage of covering fire as the JCB bull-dozed its way through the fence and into the compound. The three terrorists then jumped down from the digger's cab and ran towards the van as the large bomb was detonated in the JCB, the violent explosion destroying a large portion of the building.

The detonation was immediately followed by the SAS order to engage the enemy. The SAS and RUC opened fire. Those who survived the initial heavy volley attempted to escape in the stolen van. A hail of bullets raked through the outer shell of the van, killing its occupants in seconds. The action had taken less than a minute from start to finish, and at the end East Tyrone's ASU had ceased to exist.

The success of the SAS Regiment's ambush at Loughall clearly demonstrates the close cooperation between the various elements within the security forces in building up a picture of the enemy's intention, accurately assessing the situation, and acting accordingly. The response to terrorism must be quick and decisive, as there will often be no second chance, and while the security forces have to be selective when identifying and engaging a target, firefights with the IRA are often brutal, with no quarter given.

Other operations in the province have been less spectacular than Loughall, but have nevertheless required meticulous planning and endless patience to bring them to a successful

Below: In Ulster the SAS work closely with the intelligence agencies of the British Army.

Left: SAS soldiers frequently patrol the streets of Ulster in unmarked cars, and in plain clothes. Right: The enemy — an IRA gunman.

conclusion. In December 1983, for example, a tip-off alerted the SAS to an IRA weapons cache located in a field outside the town of Maghermulkenny, County Tyrone. An SAS unit of six troopers divided into three two-man teams and deployed around the field to wait, maintaining contact via portable radios.

One car eventually arrived and entered the field. Two armed IRA terrorists then began to recover weapons from the cache and load them into the car, whereupon the ambush was sprung. Contrary to popular opinion it is normal for SAS soldiers, circumstances permitting, to issue a warning. The SAS, like all British service personnel, operate under the Army's rules of engagement. They state that a warning should be given before opening fire to allow opponents an opportunity to surrender. However, the SAS can fire immediately if they believe the warning, or any other delay, could lead to the death or injury of themselves or any other persons, or if the giving of a warning is impracticable. At Maghermulkenny the SAS commander shouted 'Halt! Security forces'. The two terrorists then made moves which indicated that they were about to fire their weapons, whereupon the SAS opened fire with two- and four-round bursts, killing them both. The driver of the car escaped.

In urban areas SAS tactics are slightly different. Four-man teams in plain clothes frequently move around in unmarked civilian cars waiting for a call on their radio ordering them to a specified location. One such incident occurred on 2 May 1980. Eight members of the Regiment were on patrol in two unmarked cars on the streets of Belfast. They were requested to investigate a suspicious house on the Antrim Road. The lead car pulled up at the front of the house with the second car going to the rear. As soon as the SAS soldiers, dressed in anoraks and jeans and sporting fluorescent armbands for identification purposes, got out to investigate they came under fire from another house. This burst of fire killed one of the SAS soldiers, Captain Westmacott. The SAS returned fire with their personal weapons: MP5 SMGs, Colt Commandos and High Powers. The firefight resulted in the four IRA gunmen in the house giving themselves up.

The unexploded device would be left with an ambush party lying in wait

If an IRA car bomb is defused by an Army Ammunition Technical Officer or fails to go off, it can be used by the SAS as bait for a trap. The unexploded device would be left with an ambush party lying in wait around it. Invariably, the IRA will return to the bomb to either find out why it had not gone off or to recover materials. They would then be ambushed.

In Northern Ireland, as in most hostage-rescue operations, the standard use of personal communications systems has proved invaluable for CT work. The SAS can, for example, radio-in reinforcements, in addition to evacuation

helicopters and vehicles. All SAS troopers also carry 'Iris', an infra-red intruder detection system which is effective over five kilometres by remote control, and other teams can be equipped with a small three-piece ground surveillance radar with a range of 10,000m.

It is standard operating procedure (SOP) in Northern Ireland for SAS units to have a military back-up during ambush operations, either other members of the Regiment or the regular Army. In February 1985, for example, the SAS ambushed three IRA terrorists near Strabane, County Tyrone. Three troopers lay in wait in a field while three others located some way off acted as reinforcements. When contact was made the terrorists refused the SAS order to surrender and opened fire with Russian-made automatic weapons. The back-up had to be called in urgently. In a spectacular firefight the heavily armed terrorists were eventually killed; one was hit 28 times, one 13 times, with the third being struck five times.

The deadly cat-and-mouse war in Northern Ireland has also been responsible for the deaths of several of the Regiment's soldiers. In the period 1981-87, for example, the SAS

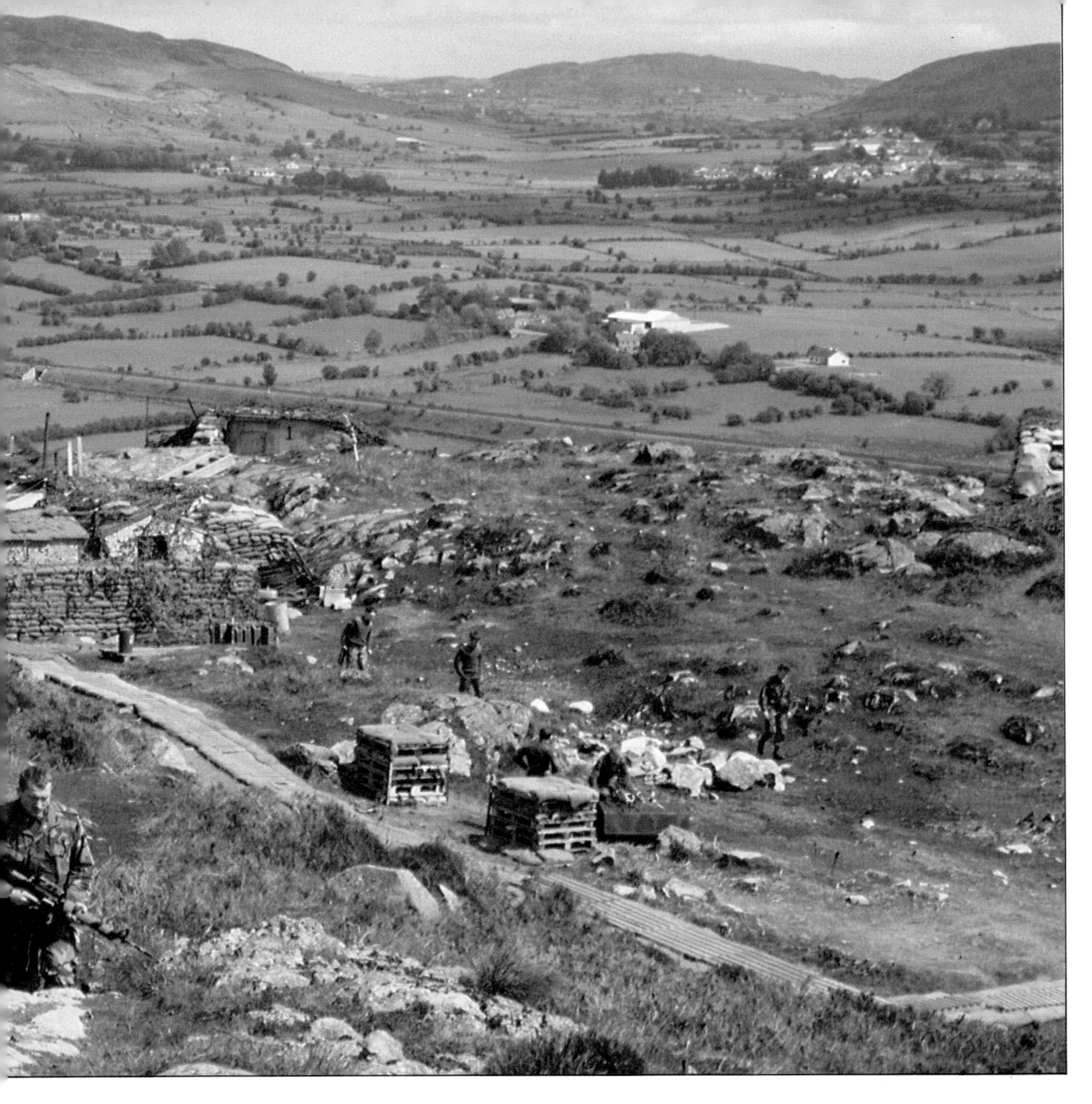

killed no less than 26 IRA terrorists in the province. However, in the same period the IRA themselves killed four SAS troopers, an extremely low rate of casualties for this type of clandestine warfare.

Whether carrying out decisive and deadly ambushes in remote rural areas of Northern Ireland, or rescuing hostages held captive by international terrorists in an urban environment, members of Britain's SAS Regiment continue to be the best in their field, and continue to refine their CT tactics, making use of the very latest technological advances in

An SAS patrol, armed with M16 rifles, on active duty in the 'bandit country' of South Armagh.

weapons and equipment. The fact that other very good CT units come to Hereford to receive SAS training is testimony to the Regiment's expertise in this particular field. One thing is certain, that while terrorism remains a threat the 'Killing House' in Hereford will be in use for many years to come. Whether at home or abroad the SAS regiment stands ready and able to deal with any terrorist attack.

ROYA AIR FORCE

FIGHTING AS A FOUR-MAN TEAM

Born in the arid wastes of north Africa in World War II, the SAS four-man patrol has proved its worth in every theatre of war, from the jungles of Borneo and Malaya to the deserts of Oman and the Gulf. Today the four-man patrol is a self-contained unit capable of undertaking a wide variety of missions.

The four-man patrol is the fundamental operational unit of the SAS Regiment. Its size and structure is an original SAS concept and it forms the basic building block of the Regiment — a small, self-contained unit within an SAS 'Sabre' (fighting) Squadron. The four-man patrol, like the SAS, owes its existence to the vision of one man: David Stirling.

In the summer of 1941, while recovering from a disastrous parachute jump, Stirling clarified his ideas on the use of small units to raid behind enemy lines. Stirling himself had participated in a number of large-scale commando raids along the Cyrenaican coast which had all ended in failure. He believed that the idea of hitting enemy rear-area supply dumps and communications was a sound one, but that the large size of the forces and equipment involved, including

A four-man patrol debussing from a Wessex helicopter in Northern Ireland. They are armed with SA-80 assault rifles, the new individual weapon of the British Army.

substantial numbers of naval vessels, compromised the element of surprise.

He therefore proposed the formation of five-man teams (not four-man) which would combine minimum manpower demands with maximum possibilities of surprise. He believed that a five-man team could in fact cover a target previously requiring over 200 commandos, and that around 200 properly selected, trained and equipped men would be able to attack around 30 different targets on the same night, as opposed to only one objective using the commando method. These teams would be trained to arrive by air, land or sea.

SAS raids in north Africa destroyed over 250 German and Italian aircraft

Following a shaky start results would prove Stirling right. Between December 1941 and December 1942 SAS raids in north Africa destroyed over 250 German and Italian aircraft alone, with enemy supply dumps and bases also being hit. The SAS went on to campaign in Italy and northwest Europe during 1943-45, but the four-man patrol had not yet achieved the high levels of skill and flexibility displayed in later campaigns.

Practice had proved that a four-man team was the most efficient and effective size for the tasks undertaken by the SAS during World War II. Four men were considered the optimum number for a variety of reasons. Less than that number was obviously too few as the patrol would be restricted in what they could carry, and limited with regard to the total amount of firepower they could call upon, as well as being unable to effectively defend themselves if ambushed or continue with their mission if one of their number was killed or wounded. Conversely, more than four men per patrol was considered unwieldly and harder to conceal from the enemy.

There were, in addition, extremely good psychological reasons for four men to make up a patrol. Members could pair up, with two looking out for each other both tactically and 'domestically', and sharing duties such as 'brewing up', cooking meals, erecting a shelter or camouflaging their position. Most men, especially soldiers, have a natural bonding instinct and divide into pairs to tackle most tasks.

The concept of the four-man patrol would be further refined and expanded during the campaign in Malaya (1952-58) by two SAS officers with the same vision as Stirling: Major 'Mad Mike' Calvert and Lieutenant Colonel J.M. Woodhouse. During World War II SAS teams had undertaken many operations behind enemy lines, liaising with, and organising, resistance groups and generally wreaking havoc. The emphasis had been on high mobility, speed, surprise and hitting power. In Malaya the four-man patrol had to be developed much further to meet the challenges of counter-insurgency (COIN) warfare. This meant each patrol member had to acquire more skills to enable him to fight and survive in the harsh jungle environment, as well as winning over a potentially hostile indigenous native population.

Conversing with the natives in their own tongue encouraged trust

As in World War II intelligence gathering was important, but now SAS patrols had to learn the art of long-range jungle patrolling to defeat the Malayan Races Liberation Army. Calvert recognised that the SAS would have to win over the indigenous aborigines in order to win the 'hearts and minds' battle of the campaign. Initially this involved such minor steps as ceasing to call them *sakai*, a Malay word meaning slaves. Much more importantly, SAS troopers were required to learn the natives' language. Conversing with the natives in their own tongue encouraged trust, but it also made the gathering of intelligence concerning the movements of the enemy much easier. This laid the foundation for a language specialist being a member of every patrol.

Calvert also worked on such things as how much ammunition a patrol should carry. It was discovered, for example, that too much ammu-

Even when 'brewing up' a patrol member never forgets the cardinal SAS rule — keep your weapon within arm's reach at all times.

nition would result in it being used wastefully, whereas much less would be used to greater effect, because soldiers were less exhausted and therefore more alert. Calvert was invalided home in 1951 but Woodhouse took up his radical ideas and continued their implementation. He developed the art of tracking so much so that the SAS became as expert as the natives. A patrol's top speed in the jungle was about 1.5 kilometres an hour, but this was discouraged as it might mean missing vital clues as to the passage of enemy forces. Therefore a slow, cautious, quiet, observant movement was encouraged.

The results were often spectacular. In the spring of 1958 the SAS carried out an operation near Telok Anson, about 45km northwest of Kuala Lumpur. D Squadron, under the able command of Major H. Thompson, was given the dangerous task of winkling out the communist guerrillas in that area, especially those under the command of the notorious Ah Hoi, nicknamed the 'Baby-Killer'. The squadron was parachuted into the area and achieved complete success with the eradication of all the terrorists, not least because of the skill displayed by individual SAS troopers in the art of tracking. Tracking in a jungle and swamp environment is an exhausting task requiring constant, non-stop, concentrated observation, made worse by the heavy rains which can destroy all traces of the trail. Ironically, the most expert SAS tracker of all was a Yorkshireman called Sergeant Turnbull. He was also fluent in Malay and had superb weapons skills. Men such as Turnbull were the epitome of the ideal SAS patrol member as envisioned by Stirling, Calvert and Woodhouse.

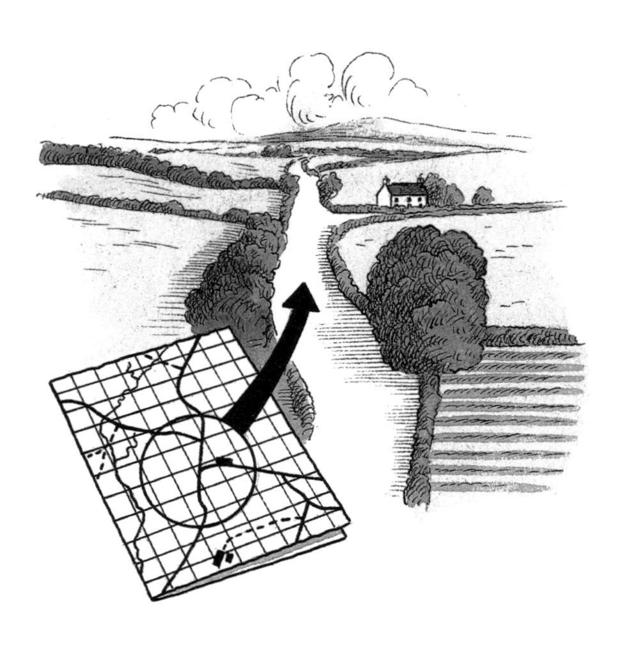

Above: Setting a map by lining up the same road on the map with the one you are standing on.

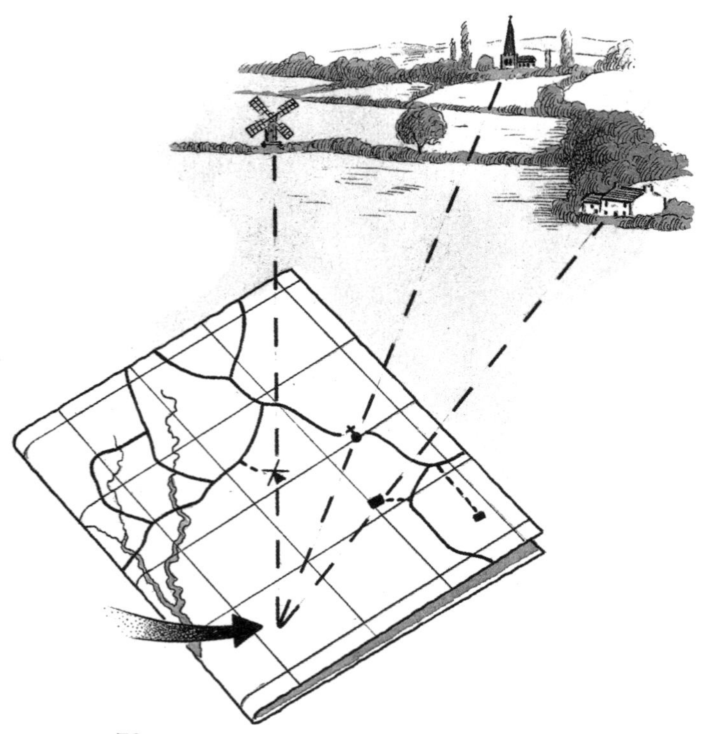

Below: Setting a map by lining up objects on the ground with the same landmarks on the map.

Calvert and Woodhouse both realised the need for supreme fitness

One aspect of the four-man patrol which cannot be overstated is the sheer mental and physical toughness of the individual patrol members themselves. Calvert and Woodhouse both realised the need for supreme physical fitness, allied to mental alertness, particularly in the draining, jungle environment of Malaya. Woodhouse especially believed that without mental stamina a man would never achieve his peak performance.

Every patrol member has his own speciality, be it signals, demolitions, medicine or languages. Communications is one of the most important skills within the SAS and all troopers are trained to Regimental Signaller standard (that which must be achieved by the communications experts attached to infantry battalion headquarters), and in morse code and ciphers. Despite this general knowledge each four-man patrol has its own specialist signaller, responsible for such jobs as calling in aerial resupply missions, casualty evacuation (CasEvac), patrol exfiltration, as well as keeping in regular contact with base. Signallers are trained in high-speed morse transmissions which can be broadcast and received more easily than voice in difficult areas for radio communications, for example when SAS teams were patrolling in the jungles of Malaya and Borneo.

Languages and demolitions skills are also important within a four-man patrol. The need for expertise in the use of explosives can be traced back to the first SAS raids against Axis airfields during World War II. Mention has already been made of SAS troopers learning the language of the natives in Malaya, where a trooper conversing with villagers could give them reassurance, in addition to gathering any information which they could provide. The same practice was undertaken in subsequent SAS campaigns, for example in Oman (1970-76) many troopers were sent on a 10-week course in colloquial Arabic before going on operational duty.

The last, and perhaps most important, patrol skill is medicine

The last, and perhaps most important, patrol skill is medicine. A patrol's medic can perform surgery and cure tropical diseases, but he is also an invaluable asset in winning the 'hearts and minds' battle. For example, in Malaya many of the aborigines had never seen a white man and were naturally suspicious and hostile. The work of the medics in treating the tuberculosis, malaria and other illnesses from which they suffered did much to win them over to the SAS. In Borneo the curing of stomach pains, toothache and other minor ailments helped form a strong bond of trust between SAS patrols and tribesmen, which paid dividends in the gathering of intelligence.

During the campaign in Oman the varied skills of the Regiment's medics became crucial in achieving victory over the guerrillas, by winning over the Dhofaris who were opposed to the Sultan's rule. Johnny Watts, the commander of 22 SAS at that time, devised a plan whereby his squadrons would provide medical services for the 50,000 Dhofaris living in the mountains. In addition, he would find vets to help the natives improve the breeding of their cattle. These methods, together with the Sultan's modernisation programme, proved to be successful and the SAS were to eventually triumph in their vicious war against the

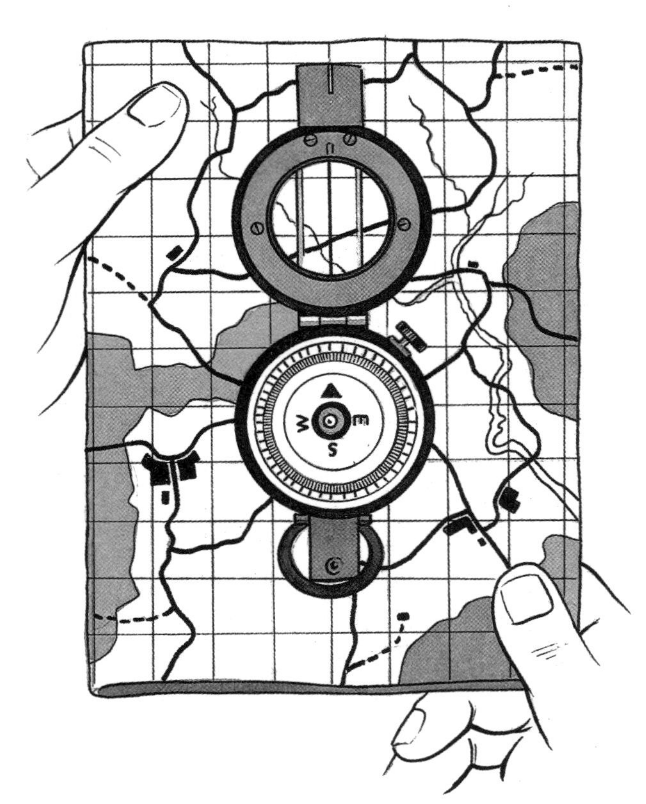

Left: Setting a map with a prismatic compass. The magnetic point on the compass must be pointing north along the north-south grid line.

People's Front for the Liberation of the Occupied Arabian Gulf.

Above: SAS patrols always maintain irregular spacing when on the move.
Left: Too bunched and exposed. One well-aimed burst would wipe out most of this patrol.
Below: Patrolling the SAS way. Each man is well-spaced to avoid unnecessary casualties.

Most members of an SAS four-man patrol have more than one patrol skill

Most members of an SAS four-man patrol often have more than one patrol skill, depending on how long they have served with the Regiment, and great importance is attached to cross-training. SAS troops, which comprise four patrols usually commanded by a captain, form the basis of the Regiment's extensive special warfare skills. Each one of 22 SAS Regiment's four 'Sabre' (fighting) Squadrons is made up of four troops, with each troop in turn having its own specialist role: Boat Troop (amphibious warfare); Mobility Troop (Land Rovers and vehicles); Air Troop (freefall parachuting); and Mountain Troop (mountaineering and winter warfare operations).

Ideally each SAS trooper is cross-trained in at least one other patrol specialisation and a second troop skill. Therefore a trooper may be a specialist linguist and medic, having trained primarily with his squadron's Boat Troop.

Thus trained he would be ideally suited to a mission which required both his personal specialist skills and his troop skill. For example, for a 'hearts and minds' operation in a coastal or riverine area he could provide medical aid to the local population in addition to utilising his language skills. Another trooper might be a signals and demolitions specialist from his squadron's Air Troop. If a situation developed that called for an attack upon an opponent's communications centre, this man's specialist skills might make him an ideal candidate for the mission. Capable of a military freefall (MFF) parachute deployment, such a soldier might be among a patrol dropped from an aircraft flying at high altitude some distance from the target. His knowledge of communications would allow him to place the charges to cause maximum damage, while his demolitions skills would be used to make up and initiate his explosives. The fact that individual patrol members receive cross-training in each other's individual specialisations and experience more than one troop skill, allows the Regiment's planners a high degree of flexibility in who is deployed and where.

Below: A needle can be magnetized to act as a compass by rubbing it on a piece of silk.

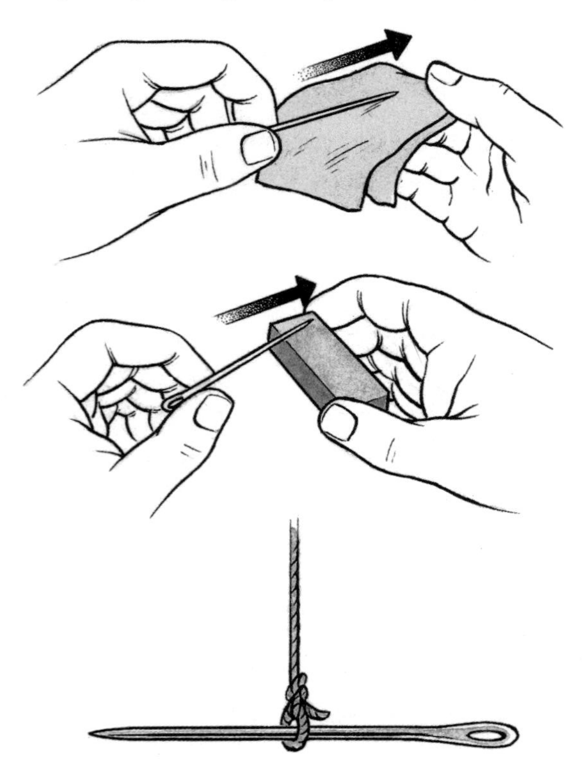

The four-man patrol is designed to operate primarily on its own, often in remote areas, and frequently behind enemy lines. In addition to general reconnaissance and patrolling, it can perform certain sabotage and ambush missions, although the four-man patrol is not in fact an aggressive fighting or combat patrol; large-scale ambushes would normally be conducted by either half or complete SAS troops (eight- to 16-men strong).

Complete, unwavering trust in each other is essential to any patrol

Basic patrol skills and techniques to the level required by the SAS can only be achieved by thorough training and extensive experience. According to the Regiment itself 'there is no such "animal" as a fully-trained SAS soldier', and neither is there any room for a man merely adequately trained. All train to a high standard in 'the pursuit of excellence' that epitomises the SAS soldier.

To be successful each patrol must comprise troopers who, if not the best of friends, at least get on well with each other and are capable of working closely together as a team. As one SAS veteran, 'Lofty' Large, states: 'Complete, unwavering trust in each other is essential to any patrol on operations'. This is especially important when one considers that patrol members might have to spend months working together, as was the case in Borneo during the British campaign against Indonesian insurgents, with little or no contact with the outside world. With the exception of the local native population, it was not unknown for a four-man patrol to be left alone and remain in the jungle for three months or more without relief.

A trooper's cross-training would allow him to take over all aspects of running a patrol, and undertake each role within it, should the situation require it. Some aspects of patrol work, such as shooting, require a high standard from all patrol members and so all are expected to be marksmen in any weapon they may be called upon to carry. Not only must the troopers in a

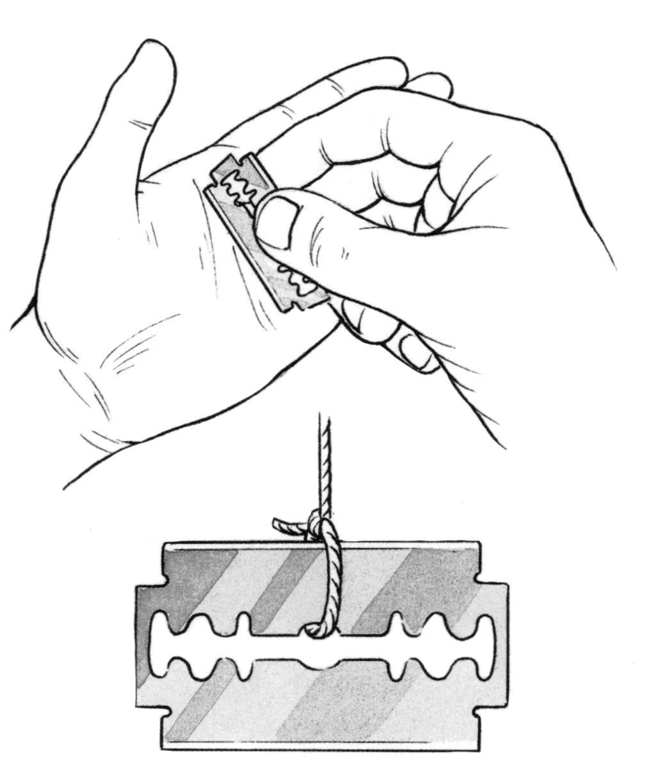

Left: A razor blade can be turned into a compass by stropping it against the palm of the hand, which results in it becoming magnetized.

four-man patrol be excellent shots from the traditional prone position, but they must be capable of sustained and accurate reactive fire at close quarters. In addition, each man must be adept at ambush fire control procedures, contact drills, fire and movement when breaking contact, and firing at night.

SAS Selection is based around timed land navigation marches

Individual navigation skills must also be of a very high standard. Using both bearing and distance methods, every member of the patrol is expected to be able to navigate using the minimum of natural and artificial aids. Compass work is important but so to is an almost instinctive ability to recognise one's position and judge distances travelled. In remote areas, especially in thick, impenetrable jungle or featureless terrain such as flat desert, map reading skills can be tested to the full, supposing that anything other than rudimentary sketch maps of such areas exist of course. SAS Selection is based around timed land navigation marches. These are designed to test stamina, endurance, fitness and determination, but without the ability to achieve a high degree of skill in navigational techniques a prospective SAS trooper is lost, figuratively and literally. All SAS soldiers are taught basic map reading skills such as setting a map by inspection (using features such as a road or other obvious land marks); by use of prismatic compass (to take bearings and triangulate); and to use compass bearings and the distance marched method (pacing) to reach a given destination either in daylight or darkness. Skills such as how to manufacture a compass from a needle or a razor blade are taught during the Continuation phase of training, but the emphasis on an indi-

Right: When marching along a compass bearing by day constantly checking bearings ensures the correct route is being followed.

JACOB

vidual's ability to navigate accurately using a prismatic compass cannot be overstated.

The military definition of patrolling 'is the means by which a force dominates an area'. Patrolling, especially on foot, is particularly important in certain environments, and forms the basis for jungle warfare — itself the *raison d'etre* of the Malayan Scouts and consequently the reason for the reforming of the SAS. Not all SAS patrols are conducted in a jungle environment, but most of the lessons learned by the Regiment during its extensive operations in the jungle since the 1950s are applicable to other theatres of operation. The jungle is a harsh and inhospitable area in which to survive and fight. It is still employed by the SAS as a final test for candidates during their Continuation training.

Reconnaissance patrols can be tasked with observation and intelligence gathering, either general or specific. This can include collecting topographical information, for instance on rivers or natural obstacles; selecting sites for rendezvous (RVs), helicopter landing sites, friendly force ambush positions or assembly positions; confirming whether the enemy occupies a particular area; or checking on friendly defences such as minefields.

Border protection is a role for which the Regiment is especially well-suited

Standing patrols, on the other hand, can consist of anything from a four-man patrol up to a troop or more, depending on their role. They can provide warning of an enemy advance as well as details of its composition; prevent enemy infiltration across borders or through gaps in the defensive system; and direct either artillery support fire or ground attack aircraft on to enemy troop concentrations attempting to cross the border.

Border protection is a role for which the Regiment is especially well-suited, and it has had considerable practical experience. In early 1963, for example, 21 patrols from A Squadron were deployed to great effect along 1000km of the Indonesian frontier to prevent enemy incursions. In the type of low-intensity operations in which the SAS specialise some border protection duties may be carried out by locally-raised irregular forces. Yet these troops may need to be supported by regular forces such as artillery and conventional infantry units and, if the conflict escalates, by engineers who can construct physical obstacles such as perimeter fences and minefields. However, if the problem requires containment rather than a show of strength or the use of force, local border guard units may simply be supplemented by SAS troops or squadrons acting as standing patrols, or patrols collecting intelligence which will provide the local military commanders with an early warning system against the build-up of enemy forces.

Patrol skills in action. A sergeant from 22 SAS and an Omani officer converse in Arabic with locals to purchase food.

The composition of a fighting patrol depends on a number of factors

Finally there are fighting patrols. These are larger in size than reconnaissance patrols and may be made up of two four-man patrols or an entire troop. The composition of a fighting patrol depends on a number of factors but most relate to its particular mission and to the type of terrain in which it will operate. The primary tasks of fighting patrols are to harass the enemy, thus lowering his morale; to conduct raids to gain intelligence and to capture prisoners for interrogation; to carry out attacks against specific targets; and to prevent the enemy from obtaining information about the disposition and strength of friendly forces operating in a particular area. In late 1964, during the campaign in Borneo, SAS patrols led large numbers of ambush parties — killer groups — over the border into Kalimantan to attack enemy forces before they could penetrate Malaysian territory.

Although aggressive operations against the enemy demand stronger and larger patrols, the four-man SAS patrol is especially useful for the more passive intelligence gathering or reconnaissance-type operations, as illustrated during the 1982 Falklands campaign. During the 1991 Gulf War small SAS reconnaissance patrols were among Allied special forces (SF) units which operated inside Kuwait, providing military commanders with vital and accurate information as to the disposition of enemy Iraqi forces, prior to the Allied land assault. Continual experience in troubled Northern Ireland keeps the Regiment's experience in reconnaissance and surveillance-type operations at an extremely high level.

The procedure for SAS reconnaissance patrols is divided into three main phases. First, on receiving orders for the particular mission the patrol conducts its rehearsal phase. Second, the actual reconnaissance patrol itself and, finally, the delivery of the patrol's report. Members of four-man patrols attach great importance to each of these phases but perhaps the preparation/rehearsal phase is the most crucial, for without this the entire operation can very easily fail disaterously.

Ideally the patrol commander (PC), usually a sergeant or corporal, should have at least a day to prepare his patrol, make his plan, brief his team and conduct a daylight and night-time rehearsal. Map information should always be supplemented, where possible, with clear aerial photographs, intelligence reports gleaned from other patrols, and a study of the ground to be covered from a forward observation post (OP). If the patrol's method of insertion is by helicopter it may be possible for the PC to fly over the area in person.

Radios are checked, ammunition cleaned and grenades primed

Rehearsals give the patrol the opportunity to sort out what equipment they need, and to make sure everything functions correctly. Radios and batteries are checked, ammunition cleaned and grenades primed, rations drawn

Below: The shaded areas represent the arc of fire of each patrol member when on the move.

The 'Tailend Charlie' regularly swings around to face the opposite direction to that in which the patrol is moving, to ensure the patrol has no blind spots.

The lead scout covers an arc immediately in front of the patrol. The second and third men cover arcs to the left and right respectively. The arc to the rear of the patrol is covered by the 'Tailend Charlie'.

(this is especially important for lengthy, long-range reconnaissance patrols), water-bottles filled, and weapons test-fired.

Each patrol equips itself according to the task in hand and, although there is a degree of flexibility in what is carried and what is left behind, there are some basic ground rules for all missions. Each member of a patrol is responsible for his own personal and specialist kit, while the commander is responsible for checking each man's individual equipment and dividing the specific-to-task kit throughout the patrol, dividing the loads and ensuring that each man knows what the others are carrying.

An individual's equipment check-list would include personal weapon, bergen and belt kit. Spare ammunition for his personal weapon (apart from the magazine loaded in the weapon itself) would normally be carried either in pouches on his belt kit or in chest webbing (an idea copied from communist insurgent forces in the Far East). Extra ammunition would be carried in a man's bergen, either in 'link' belts for the GPMG (if carried), or loaded into spare magazines. Each man's personal belt kit would contain his additional survival and medical gear, fragmentation and smoke grenades, water bottles and emergency rations. The bergen carries everything else: radios and batteries, ammunition and explosives, rations and water, sleeping bag and spare clothing.

The PC and patrol second-in-command (2IC) would usually carry items specific to their own role within the patrol. The PC's load would normally include additional command and navigational equipment, including more detailed maps; passive night vision goggles (PNGs) and a hand-held thermal imaging device; a spare short-range radio; and a Sarbe (search and rescue beacon) for communication with support/extraction aircraft.

Below: Winning 'hearts and minds'. An SAS medic treating local tribesmen in Oman.

The lead scout is responsible for taking the 'point', i.e. marching at the front of the patrol's line-of-advance. He will normally carry additional night-vision aids; wire cutters and hessian for dealing with barbed-wire fences; equipment for dealing with mines and booby-traps; and perhaps an M203 grenade launcher to put down suppressive fire in the event of a contact.

Every patrol member must be as switched on as the leading scout

The patrol's order of march depends on such factors as the tactical situation and the terrain to be crossed, but usually comprises lead scout followed by PC and signaller, with the 2IC as the 'Tailend Charlie' bringing up the rear. Three types of patrol formation are common for a four-man patrol: file, single file and diamond. File (or rectangular) formation is suitable for following linear features at night, with the PC and lead scout acting as one pair and the 2IC and signaller as another. Single file formation is the most common with the spacing dependent on the type of terrain being crossed. It is especially useful in close-country or jungle but has the disadvantage of being very vulnerable to fire from the front. Lastly, the diamond formation might be used, especially in open ground at night. This combines some of the best features of both file and single file formations, allows optimum firepower to be focused to the front, but is less suitable for close-country or daytime use. 'Lofty' Large describes the tasks of the patrol members thus:

'The leading scout has the job of finding the way through, as quietly as possible, while still navigating to maintain overall movement in the right direction. At the same time he has to be ready to react like lightning. Every patrol member must be as switched on as the leading scout. Their extra problems are that their friends are probably between them and the enemy. They must be ready to bring instant fire to bear in the right places while being careful not to hit their own men.'

'They must keep their eyes open to left and right for any sign of problems, and make sure the man behind is still where he should be. They must keep in sight of the man in front but must be far enough back not to be caught in the same burst of automatic fire, mine or booby trap. In a contact with the enemy they must react instantly, guided by the actions of the man in front of them, or what happens to him.'

Rehearsals, if time allows, are practised with the emphasis placed on actions and drills to be carried out in case of contact with the enemy, on moving into the final RV, and on reaching the objective. Also important are the order of march and what signals will be used on changing formation and drills to be carried out on encountering obstacles both natural and man-made. Other tactical procedures that will be discussed within the patrol, if not actually rehearsed, are the actions taken on encountering mines or flares, receiving a casualty and capturing an enemy prisoner. Procedures to be taken on, and during, patrol halts will also be practised during the rehearsal phase.

The head-on contact drill is one of the most effective four-man patrol tactics

In a four-man patrol procedures need to be 'slick' and rapid. Tactical advantage must be maintained through use of terrain with two men remaining alert at all times, allowing the remaining pair to drink, eat, check weapons and re-pack their kit. Then the positions will be reversed. On extended patrols the men will find a suitable location to adopt as their lying-up position (LUP). Depending on the tactical situation, as well as the terrain, the patrol may move into an LUP either during the day or at night. Because of its small size the four-man patrol may not always detail a sentry but might in some cases rely on camouflage and concealment for its defence, but normally one man at least will remain alert and awake.

The head-on contact drill is one of the most effective tactics of the SAS four-man patrol. If contact is made with the enemy the patrol members will instantly manoeuvre to place themselves in a position whereby they can fire without hitting their comrades. For example, if a patrol is in file formation and the lead scout makes contact with hostiles, the

SAS patrols have devastating firepower.

other three troopers will break left and right, bringing their weapons to bear on the enemy. After the initial contact, if the enemy is largely destroyed, the patrol may wish to advance and finish them off. However, should the patrol wish to withdraw then two will lay down covering fire while the other two withdraw a distance. The latter will then be in a covering position, allowing the first pair to fall back and so on.

SAS fast reactions and rapid firing techniques can sometimes lead to unexpected problems. During a four-man operation against Indonesian shipping on the Koemba River, Borneo, in 1965 'Lofty' Large's patrol were retreating after a highly successful ambush of an enemy launch. Moving rapidly through the jungle Large was stopped in his tracks by a huge king cobra snake, which reared up in front of him, however he dare not shoot: 'A shot would not only give us away to the enemy, but would put the patrol into its head-on contact drill. In other words create a small war on a snake.' The cobra eventually bolted, leaving Large to complete his speedy withdrawal.

On a reconnaissance patrol contacts with the enemy should be avoided at all costs, shooting at targets of opportunity is definitely out and a four-man SAS reconnaissance patrol will only use its weapons as a last resort, i.e. to break contact with the enemy and escape. The patrol's primary mission is only over when the information it has gathered has been successfully transmitted back to its base, normally by radio, and supported by a written report and verbal debriefing.

The four-man patrol was designed to be a self-contained unit. Like the Regiment itself, David Stirling's concept has withstood the test of time. SAS patrols were as effective in the 1991 Gulf War as they were against Rommel's forces in the desert of north Africa during World War II. There can be little doubt that the four-man team will continue to be at the heart of the SAS operational art of war for the foreseeable future.

SILENT KILLING SKILLS

On clandestine operations behind enemy lines the ability to silently kill enemy personnel can mean the difference between a mission's success or its failure. As a result, SAS soldiers are highly trained in the art of silent killing and the use of a wide variety of weapons, ranging from machetes, spears, suppressed firearms and knives to state-of-the-art compound bows and crossbows.

Silent killing techniques are taught to a select number of soldiers, including those experts belonging to special forces (SF), airborne troops and commandos. They also play a small, though vitally important, part in the Continuation and follow-on training of SAS troopers.

Silent killing techniques can be especially useful during night operations, where darkness allows the trooper to close with the enemy without being seen. One by-product of these skills is the effect on the morale of the enemy caused by silently eliminating his forces while remaining undetected. Such techniques were used against members of Iraqi beach patrols and installation guard forces by swimmer/canoeists of the Special Boat Sections. These elite specialist units of the Royal Marines Special Boat Squadron (SBS) are similar to SAS Boat Troops, and were employed extensively during the 1991 Gulf War to reconnoitre possible amphibious landing areas along the Kuwaiti coastline.

A rifle fitted with a bayonet can be used as a savage stabbing weapon for hand-to-hand combat. The weight of the rifle and the power of the strike ensure that the first blow is fatal.

Right: The modern crossbow is accurate, lightweight and easy to assemble and use.

Trained in the same techniques as SAS troopers, the deadly SBS commandos killed enemy soldiers as a part of the overall Allied strategy to convince Iraqi commanders that an amphibious assault from the sea would form a major part of the Allied attempt to retake occupied Kuwait. By killing a few enemy personnel and by leaving tell-tale signs of their nocturnal visits, Allied SF soldiers made sure that the Iraqi high command committed a substantial number of its forces to the guarding of the Kuwaiti coastline against an invasion which never came.

The techniques used in the art of silent killing are necessarily very simple. Those men having to employ them will probably be tired, as they will have most likely been engaged on intensive operations for some time, and under conditions of extreme stress, leading right up to the moment they actually apply the techniques learned during training.

Most of the silent killing techniques taught to the modern SF soldier originate from those developed by Allied commandos during World War II. Using a combination of various martial arts skills, coupled with a basic common sense approach and a working knowledge of human anatomy, various methods of 'taking out' an enemy soldier as silently and safely as possible were devised, tested, and either discarded or adopted. The idea then, and this remains true to this day, was to train all special operations personnel to such a high level that they would react instinctively.

The human body has numerous vulnerable organs and areas which, if destroyed, will result in death. These include the heart, lungs, liver, neck and the central nervous system. For silent killing techniques the neck and heart are perhaps the most accessible targets. The neck contains the spinal column and the jugular vein. Break it with a martial arts blow or hold, or with a strike from a weapon or blunt instrument, and the victim dies without uttering a sound. Cut the jugular and the victim drowns in his own blood, the only sound uttered being a low gurgle (though care must be taken as his body may go into spasms and start thrashing

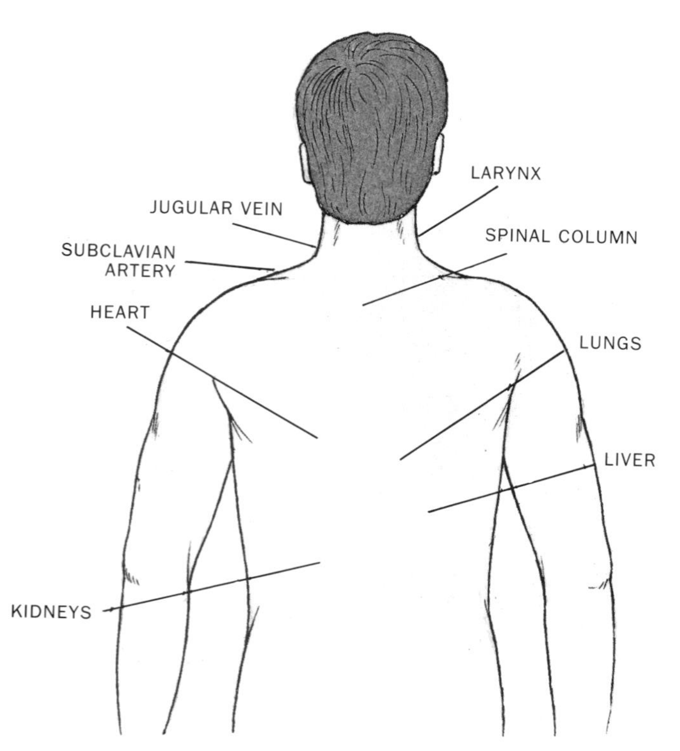

Left: Rear view of the human torso indicating the main targets areas. In training SAS troopers are instructed to aim their blows at these targets.

Left: The compound bow. The pulleys fitted to each end of the limbs increase the arrow's speed.

around). Similarly, the heart is also a prime target, though any type of body armour, backpack or webbing can make it difficult to hit.

To strike such targets the SAS trooper must get within reach of his opponent. To do so he employs fieldcraft techniques such as stalking, developed over the centuries by men who hunted game in order to survive, and which has been adapted to deal with a larger and more deadly prey, i.e. man. Enemy sentries were, and remain, the most obvious targets for the silent killing skills of the SF soldier. Many of the techniques employed to deal with sentries today, such as always approaching the target from the rear or from the sides, but never from the front, have been tried and tested for years and remain unchanged.

Modern silent weapons, such as the crossbow and lightweight compound bow, mean that enemy sentries and guards can be killed silently, and at distances of up to 100m and beyond. The lightweight, slim alloy bolts and arrows that these weapons fire can go through clothing, webbing, even modern armour vests, and still inflict a fatal wound. Their one drawback is that they require a high degree of skill to use effectively, as the first shot must always hit its target.

Approach from behind and attempt to cover the sentry's mouth

Other silent killing techniques involve the use of silenced or suppressed firearms. With a noiseless or very quiet weapon, such as a silenced submachine gun or semi-automatic pistol or revolver, the need to close with the enemy in order to kill him has lessened quite considerably, though handguns should be used at close quarters to be effective. The one major disadvantage with silenced weapons, especially pistols, is that they are not as accurate as their unsilenced counterparts.

Right: Spears can easily be made from tree limbs. The thrower is used by the Australian Aborigines.

Left: Silenced weapons. From left to right: Ingram submachine gun, Sterling submachine gun, Parker-Hale sniping rifle and Beretta 92F handgun

There will undoubtedly be situations in the future, just as there have been in the past, where an SAS trooper will have to use more traditional methods, such as a knife rather than a suppressed firearm, to silence an enemy soldier. The standard technique is as follows: approach from behind and attempt to cover the sentry's mouth with one hand while using the knife with the other. The target area depends on a number of factors, such as whether the sentry is wearing webbing which will prevent stabbing the back, but one of the more common areas is the throat. Other important aspects of the attack that should be considered include where the body is going to fall after the strike. If it falls off a wall or roof, or into a river, it will probably result in a loud noise which might attract attention and alert other guards.

The advantages of many different types of weapon, if there is a choice, should also be considered. Using pointed weapons, such as a dagger, as opposed to those with a sharp edge, is generally preferred as the sentry should be stabbed rather than slashed. However daggers, such as the Fairburn-Sykes commando knife or the German Army boot-knife, are not always available, while other with more general applications are. Survival knives, for example the US aircrew survival knife, are standard issue. However, this particular model has a serrated upper edge, making it unsuitable for stabbing as the teeth can become caught in clothing.

Home-made spears can be used for hunting fish and game, or man

Other sharp-edged instruments, such as *goloks* and machetes used by the SAS for clearing areas of dense jungle and for chopping wood, are standard-issue items which can also have a secondary application. Their primary role is for heavy use against inanimate objects, but in

Right: Edged weapons. Survival knife (extreme left), machete, axe and dagger.

certain situations they can, and no doubt have, been successfully used as vicious anti-personnel weapons. They are essentially slashing, as opposed to stabbing or thrusting, weapons, and as such cannot be guaranteed to kill with one blow. However, a well-aimed strike with a sharp blade to the side of the neck could, in theory, decapitate an opponent. It is reported that a severed head remains conscious for up to twenty seconds after being detached from the body. SAS personnel need not worry, however, there are no records of any such heads speaking after separation!

Other ad hoc weapons include such things as axes (including arctic-issue ice axes), shovels and entrenching tools. Improvised weapons can also be used, although these are normally reserved for those men operating outside the sphere of conventional SF missions. However, in a combat survival or escape-and-evasion scenario the ability to manufacture weapons can mean the difference between life and death, and all SAS troopers are trained in these skills during the Continuation phase of their training. Construction of elementary weapons such as spears, using a wooden staff or pole, and either a sharpened bone or knife blade, or some similar object as a point, are skills taught to those attending SAS jungle survival courses. Home-made spears can be used for hunting fish and game, or man.

Below: Improvised weapons. Home-made bow, a 'blackjack' (a sock filled with sand), a sharpened wooden stake and a garrotte made from wire.

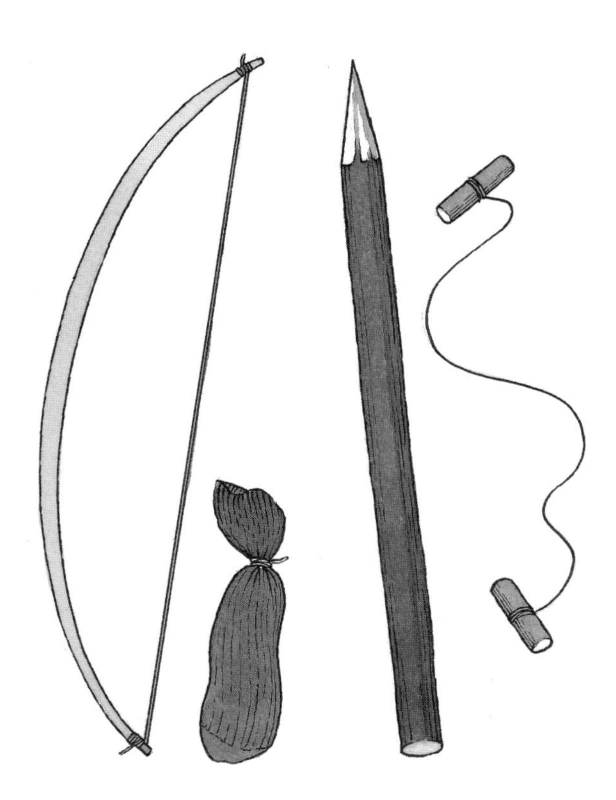

Garrotting is one of the simplest methods of killing an enemy

Apart from improvised weapons SAS soldiers, like all personnel belonging to the combat arms, are issued with a bayonet. Intended for use with a rifle as a close-quarter weapon, the bayonet can be employed on its own as a knife or dagger (a dagger is a short stabbing weapon with a pointed blade, a knife is a cutting weapon with a sharp-edged blade).

Garrotting is, perhaps, one of the simplest, yet effective, methods of silently dispatching an enemy. A wire is quickly wrapped around the victim's neck and both ends are then pulled tightly in opposite directions — death comes speedily. Almost any type of cord, wire, even vine, can be used, though the connoisseur will employ a length of razor wire.

Most of the techniques employed to silence sentries are taught as a part of the SAS Regiment's unarmed combat course, and during regular refresher courses. A high standard of self-defence and hand-to-hand combat skills are expected of each SAS trooper, regardless of whether they will have to use them or not. Although the men themselves are not specially selected for their violent nature, the SAS know that in order to survive either in a close quarter combat situation, or on a special mission where it is likely that a trooper will have to silently kill an enemy in 'cold blood', they need a different type of soldier, one who can switch off on an emotional level and let instinct and training take over, should the situation demand it. One thing is certain, while SAS soldiers have to operate behind enemy lines, the techniques of silent killing will still be taught at Stirling Lines in Hereford.

MEDICAL EXPERTISE

Medical skills form an essential part of an SAS soldier's training. Rather than having to depend on modern, sophisticated drugs, the Regiment prefers to rely on simple techniques for curing even the most serious illnesses and wounds. There is a very good reason for this: patrols, often operating far from friendly hospital facilities, have only their own medical packs and locally-available materials to serve their requirements.

The importance of medical skills to an SAS patrol cannot be over-stressed. Operating as they do in small groups, often far removed from areas of civilisation and outside medical assistance, and spending long periods working within hostile environments, SAS soldiers must be self-sufficient in all respects.

During World War II the first SAS troops had only the minimum of medical knowledge or skills. The four-man SAS team devised by David Stirling, the Regiment's founder, did not at first contain the now well-known specialities of signaller, linguist, demolitions expert and medic. Unlike conventional combat units such as infantry or armoured cavalry, the wartime SAS patrols were not in the front line: they operated beyond it. After action every effort was made to recover friendly casualties, often at great personal risk to the rescuers, but these men had little chance of surviving the long journey by vehicle back to

Learning medicine the SAS way. Clean out the wound with hot water if available, if not use urine; dig out all bits of metal and clothing; if blood is spurting out plug the hole with your finger.

THE KISS OF LIFE

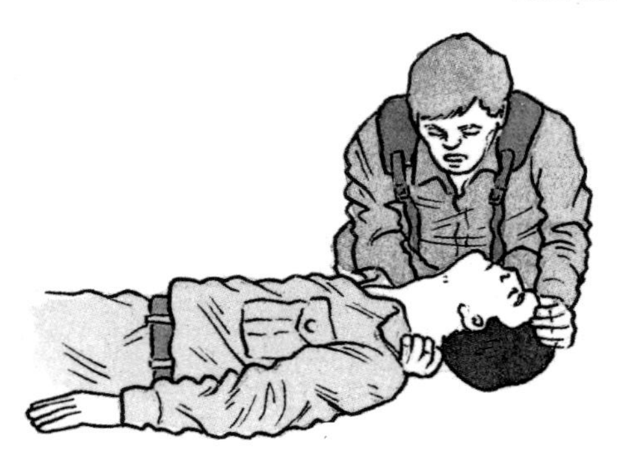

1. Place casualty on his back, loosen his clothing, lift his neck slightly with one hand and, with the other on his forehead, tilt his head backwards.

2. Remove hand from casualty's neck. Close and seal his mouth with free hand and seal mouth over nose. Take a deep breath and blow hard enough to make his chest rise. Repeat every six seconds.

CLEARING AN OBSTRUCTED AIRWAY

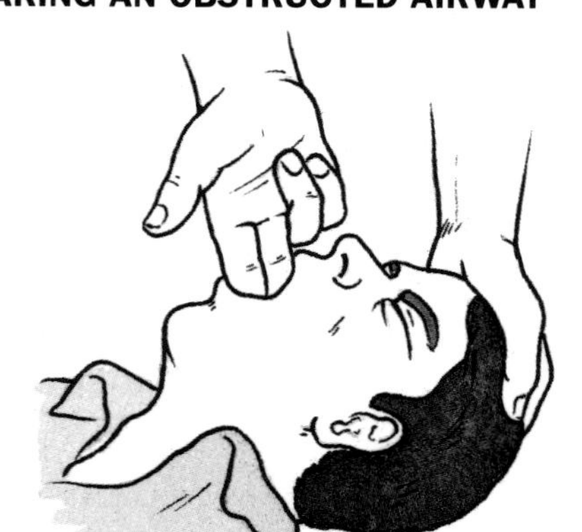

Remove anything tight from around casualty's neck. Remove all foreign material from mouth and throat with fingers.

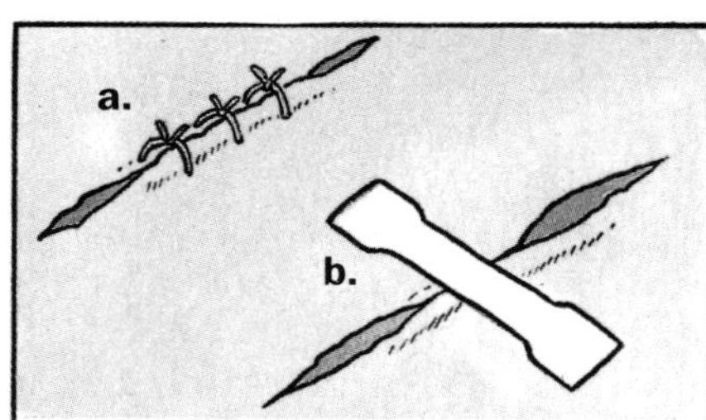

a. *Stitches.* With needle and thread make each stitch individually.

b. *Adhesive sutures.* Draw edges of wound together and apply adhesive plaster.

DRESSING A WOUND

1. Open field dressing and press it into the wound. Cover wound completely.

2. Bind the dressing firmly into place.

3. Place casualty's forearm across his body with his elbow at right angles.

friendly lines, or to hidden bases behind enemy lines, unless they received immediate and expert skilled medical attention. As the war progressed, more and more experience in dealing with casualties was gained, and SAS troopers began to be instructed in basic and advanced first aid and casualty care.

Successive post-war campaigns, first in Malaya and then in Borneo and Oman, brought Britain's SAS Regiment more and more practical experience in the art of first aid, the care of casualties, preventive medicine, and the treatment of tropical and exotic diseases. One asset available to the SAS during the latter stages of the Malayan campaign, and throughout the Borneo confrontation, was the helicopter. Unlike SAS units operating in the Western Desert during World War II, members of the Regiment operating in the jungle less than 20 years later could often rely on evacuation by helicopter in the event of an emergency. SAS casualty evacuation (CasEvac) by means of heli-

Left: On operations, basic medical skills and rapid action are often enough to ensure that the casualty has a chance of recovering from his wounds.

Above: An American CasEvac operation underway during the 1991 Gulf War. As the SAS discovered in Malaya, the helicopter is ideal for this role.

copter ensures prompt specialised medical attention and is used for serious casualties whenever the tactical situation allows. Urgent cases include casualties with major haemorrhages, severe burns, multiple wounds or serious wounds to the chest or abdomen, major fractures, severe head injuries, or hypothermia.

There are two basic medical kits carried by SAS patrols while on operations: individual medical packs and patrol medical packs. Each SAS trooper carries an individual medical kit either in his webbing or on his person. This kit comprises the essentials for immediate first aid and the treatment of most common medical complaints. This varies according to the area of operations and the specific mission, but usually contains the following items: syrettes of morphine or similar (strong painkiller); Gentian Violet (mild antiseptic); Chalk and Opium (for diarrhoea and similar intestinal disorders); Neomycin Sulphate (strong antiseptic cream); Tetracycline (broad spectrum

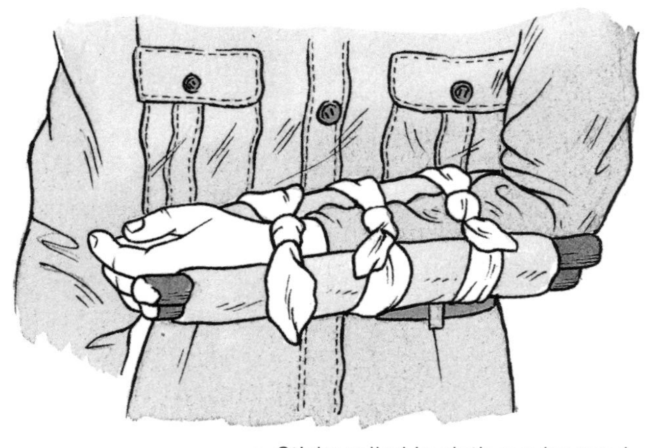

1. Sticks rolled in cloth can be used to form an improvised splint for broken bones in the arm.

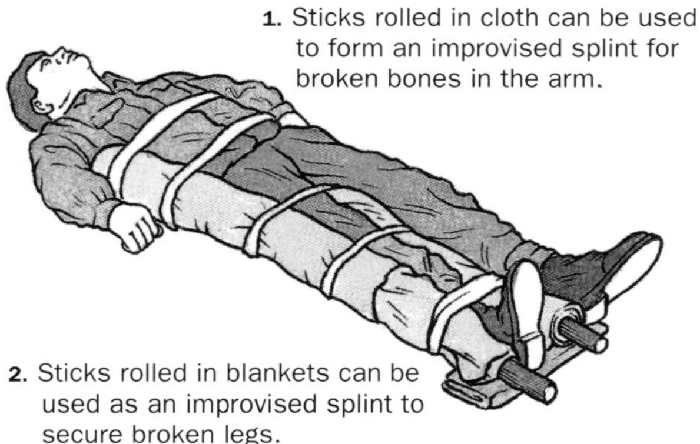

2. Sticks rolled in blankets can be used as an improvised splint to secure broken legs.

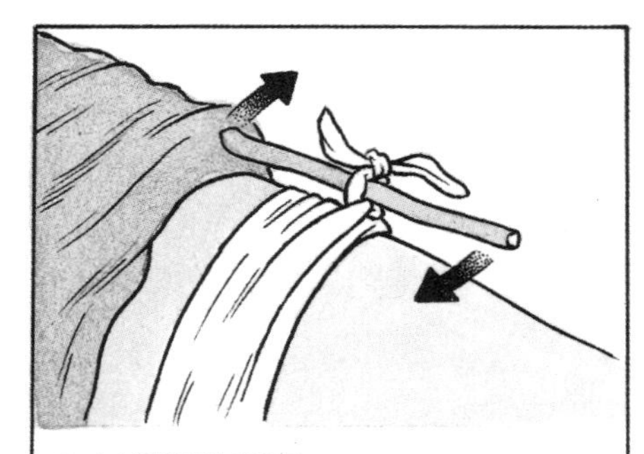

3. A TOURNIQUET
Cloth is wrapped around limb three times and tied with a half knot. A stick is placed over knot and then secured with a double knot. The stick can be twisted until bleeding stops. Wound can then be worked upon quickly. After job is finished release completely. Tourniquets should never be applied on the head, neck or torso.

Above: The ability to improvise is impressed on all SAS soldiers. Most environments will yield materials which can be used for medical treatment.

antibiotic); Codeine or similar (orally administered pain killer); dressings and assorted plasters. In addition, each man always carries at least two wound dressings on his webbing or in his clothing, and would probably carry a couple of morphine syrettes attached to his identity discs while on operations.

Patrol medical packs are far more comprehensive than those carried by individual SAS troopers, and contain a greater selection of medicine, dressings and drugs, as well as a surgical packs and a dental repair kit. Like the kits carried by all troopers, the patrol medical pack is always carried in a waterproof container complete with instructions for use. Extra items may be added by the patrol medic and these would no doubt include treatments for local diseases (especially in the tropics), an important consideration in patrols of long duration.

The SAS skill of improvisation is readily apparent in the area of medical specialisation, and SAS-trained medics are not above adapting local treatments or techniques, if they are effective. Making splints and stretchers from locally-available materials, for example, is an especially useful technique when operating in very remote areas, and such skills are taught to all SAS personnel during the Jungle Phase of their initial training. Jungle litters, for example, can be manufactured from bamboo or saplings bound together with parachute cord or creepers, and suspended beneath a long pole to form an improvised stretcher for evacuating sick and wounded men.

All SAS troopers receive intensive instruction in the treatment of various wounds during their initial training. The need for prompt action in the treatment of wounds and illnesses is impressed upon all troopers. It is a simple truth that many soldiers have been saved from death or permanent disability simply because immediate emergency care was rendered. On long-range patrols deep inside enemy territory, and away from hospital facilities, the SAS team often have no option but to rely on uncomplicated solutions to their medical problems. However, primitive techniques are often the most effective. The basic principles of treating wounds are simple: stop the bleeding; ensure airway is open; protect the wound with a dressing; apply a padded splint to any broken bones, to include the joint above and the joint below; and position the casualty to the best advantage.

Maggots can be used to treat infected wounds as they eat only dead tissue

Loss of blood is a major cause of fatalities in war, either through direct bleeding to death or as a result of shock. However, a snug dressing, if applied in time, can control around 95 per cent of all wounds. If not, then a tourniquet will be used by the patrol medic, though in hot climates there is only a short time for any operation to take place before the flesh below the tourniquet dies. If the wound becomes infected the medic will make one or more cuts in the pus pockets to allow them to drain. The application of mild heat on the wound several times a day, such as warm canteens or hot wet cloths, improves circulation and encourages an infection to localise towards the surface.

The treatment for burns is similarly rudimentary. The burned area has to be washed out and then covered with a sterile dressing. If no water is available then urine, an extremely sterile liquid, can be used instead. In addition, maggots can be used to treat infected wounds as they eat only dead tissue. Finding these creatures is easy; in any warm climates the wound has only to be exposed for a short time, the maggots will find it.

The SAS believes that prevention is better than cure, and so all troopers are well-versed in preventive medicine. As a matter of course all soldiers receive immunisation shots. Where possible, all drinking water is purified with iodine or chlorine powder to prevent the contraction of intestinal diseases such as cholera, diarrhoea and typhoid fever. Tea contains tannin, and so 'brewing up' can help to prevent dysentery. Worms — round, hook and tape — are a constant hazard while on missions, either from food eaten or filthy living conditions. Personal hygiene is the best preventive measure, but swallowing two tablespoons of kerosene or gasoline, if available, will flush any worms out of the system.

Insect-borne diseases are a particular danger which soldiers must guard against, especially in desert and tropical regions. Insects can introduce numerous diseases through bites to the skin including malaria, yellow fever and dengue from mosquitoes; typhus and trench fever from lice; spotted fever from ticks; scabies from mites; scrub typhus from chiggers; plague and murrain typhus from fleas; and sand fly fever from sand flies. Preventive measures include using insect repellent and insecticides to keep the insects away. An effective method of preventing malaria, the most prevalent insect-borne disease in the world, is to take a suppressive drug, such as chloroquine, once weekly to prevent the onset of the disease.

SAS soldiers have been known to stuff open chest wounds with handkerchiefs

SAS soldiers specialising in medical skills are also given the chance to attend the US Army's special forces medical course. Recognised the world over as the most intensive course of its kind, it is based at Fort Sam Houston, Texas, and at the special forces headquarters at Fort Bragg, North Carolina. It lasts just less than a year and qualifies those who pass as 'paramedics' — 'just the type of people you want around when your life hangs by a thread', according to one SAS source.

One further factor that must be mentioned is the sheer mental and physical toughness of SAS troopers. The 'will to win at any cost' mentality is important in overcoming wounds and illnesses, and greatly helps the work of the SAS medics. A positive mental attitude often helps the healing process far more than the most sophisticated of modern drugs. SAS soldiers have been known to stuff open chest wounds with handkerchiefs and torn shirts, to have amputated a limb suffering from a severe wound, and keep going. All things are possible for those with the will and determination.

CIVIL DISORDER CONTROL

In the event of armed extremists using the cover of widespread civil disturbances to further their own ends, the SAS can be called in to 'surgically remove' these subversives. Though the Regiment has the skills for dealing with urban combat, its use against armed civilians would only take place if the government was facing a real threat to democracy.

Any democratic government must guard against the break down of law and order, attempts to seize power by extremists, and the possible burning and looting of its major towns and cities by violent mobs. The British government is no exception in this. Though the civil police force in Britain is capable of dealing with most cases of public disorder, the SAS are trained in the tactics of counter-subversion. In the unlikely event of large-scale civil unrest on the British mainland and in Northern Ireland, SAS teams would be used to prevent bloodshed and destruction, though the extent of their involvement in civilian disturbances is extremely difficult to determine.

Its ability to be used as a strategic 'tool' of government policy is not one readily admitted to by either the Regiment or its political masters, but it remains ideally suited to take on a wide range of clandestine, non-attributable actions, should the need arise.

Scenes such as these from Northern Ireland are a constant reminder of the need to have contingency plans in the event of similar large-scale violence in the cities on the mainland.

SAS expertise and experience in hostage-rescue operations obviously make its personnel an important asset when it comes to advising other recognised authorities involved in ending sieges of this nature, especially when the lives of many hostages are at risk. It is known, although officially 'unconfirmed', that SAS advisors have been present at a number of prison riots in recent years. In October 1987, for example, SAS troopers abseiled down the walls of Peterhead Prison in Scotland, and stormed the building to free a prison officer who was being held by three inmates.

The snipers would be used to provide accurate long-range fire

Britain's SAS Regiment has a well-deserved reputation for planning and preparing for almost any eventuality, often before a particular type of threat is readily apparent. The decision to use SAS teams in any given situation depends on the prime minister, who is advised on the capabilities of the SAS by senior security advisors, including a regimental liaison officer. In the event of the break down of law and order within the United Kingdom, it is likely that SAS units would be available to deal with certain aspects of civil disorder in the same way as they might respond to a specific terrorist threat. Though such a situation may appear unlikely, the riots that occurred in some of Britain's major cities in the summer of 1981 are an indication of how civil disturbances can often quickly erupt. One alarming aspect of these disturbances was that certain areas, notably Handsworth in Birmingham, became police no-go areas for a short period. During this time mob rule became the order of the day.

The SAS does not have the manpower to maintain cells in every major town and city for the purposes of counter-subversion. However, any trouble spots can be reached quickly by means of either helicopter or aircraft. In any civil disorder the local police force will initially have control over the situation. The SAS have no specific role as such, unless given one by order of the Home Secretary. However, the police can pass control to the SAS should the need arise, as at the famous siege of the Iranian Embassy in the summer of 1980.

Above: An SAS sniper, acting under the orders of a senior civilian police commander, neutralises an armed *agent provocateur* during a violent inner-city riot on the British mainland. In such a situation good radio communications are vital.

In such cases SAS personnel would be deployed as snipers and snatch teams to pick-off armed violent extremist agitators during the course of large-scale violent riots, where police resources are stretched to breaking point. In a 'worst-case' scenario, armed agitators and political extremists might use the riot as a cover for killing policemen, arson, looting, and generally furthering their own aims by escalating the situation.

Small numbers of highly trained SAS snipers, used in support of units on the ground, could be used to contain any armed insurgents, intent on killing policemen, who were dispersed among the rioting crowds. The snipers would be used to provide accurate long-range fire from positions which allowed continuous support to police units. In such situations the rapid exchange of intelligence is of paramount importance. To facilitate this all snipers would be equipped with personal radios, allowing them to communicate with police officers on the street, commanders located in rear-area operations centres, as well as observation helicopters and aircraft operating in the vicinity.

In towns and cities sniper fields of fire and observation areas are defined by streets and major roads. The sniper must choose a position which will give him good observation, good fields of fire, and also the maximum amount of camouflage and cover. In addition, snipers should, if possible, be positioned in buildings built of brick or stone; buildings constructed of inflammable materials can catch fire, or be set on fire by an angry mob, thus rendering the sniper position untenable.

The last thing a rampaging mob is thinking about is a hidden sniper

SAS skill in concealment and camouflage can be utilised to maximum effect during a civil disturbance. A sniper occupying a room, for example, will fire from a position back in the room. The sound of the shot will be muffled and the muzzle flash will not be noticed. The last thing a rampaging mob is thinking about is a hidden sniper located in excess of 500m away on a roof top. The psychological effect of an unexpected, well-placed shot can be considerable. It will often result in the crowd rapidly dispersing, if only temporarily, allowing the police to seize the advantage.

Most snipers will be employed in support of units on the ground, be they police teams, firefighters, or ambulance men. However, there will also be snipers operating as independent teams — hunter/killer squads — whose primary task will be to search for targets of opportunity, notably armed insurgents and their leaders.

In addition, the need for good, accurate and up-to-date intelligence concerning the presence of agitators and activists is another aspect of SAS operations which would be of considerable value to the local authorities in such situations. In this instance SAS soldiers would operate in plain clothes, collecting valuable intelligence.

Apart from acquiring skills which may someday be used to restore order, and constantly maintaining their CRW and conventional special forces skills, SAS soldiers are also involved in diplomatic protection and providing bodyguards for government officials involved in overseas visits. The intense training given to SAS recruits in close quarter battle skills make them ideal candidates for protective assignments. Their skill in being able to rapidly draw handguns and submachine guns from concealed holsters to kill an assassin before he can fire, is an invaluable asset for bodyguard work.

Above: A sniper from the Regiment covers firefighters and police during large-scale disturbances. Note the radio contact between him and the police commander on the ground.

In the SAS the British government has a unit which can be used on the streets of its towns and cities, should the need arise. The Regiment's previous experience with the police during such incidents as the Iranian Embassy siege and numerous prison riots, has resulted in it having a good relationship with a number of police forces in Britain. The SAS is well-equipped to deal with armed aggression on the streets of the United Kingdom.

OBSERVING THE ENEMY

Excellent camouflage and concealment skills, coupled with sheer daring, have resulted in SAS observation techniques being the best in the world. From the cover of cramped, uncomfortable hides, SAS teams are trained to relay valuable intelligence concerning enemy activity back to headquarters. From the 'bandit country' of South Armagh to the ice-cold wastes of the Falklands and the deserts of the Gulf, the SAS Regiment has proved itself to be one of the most audacious military units in the world with regard to establishing covert observation posts.

The gathering of military intelligence, either from static observation posts (OPs) or mobile reconnaissance patrols, is a major part of the Regiment's wartime role, and SAS troopers are expert in the selection, construction and manning of covert OPs behind enemy lines. From the jungles of Borneo to the hedgerows of South Armagh, many SAS troopers have experienced the cramped confines of covert OPs or 'hides', as they are also known within the Regiment, often for extremely lengthy periods of time. Although covert OPs can be established in urban areas, most are set up in rural areas well away from towns and cities.

Two members of a patrol observe enemy activity. The main part of the Regiment's wartime role consists of collecting information on an opponent's military capabilities.

The most important aspect of any OP work is the selection of a suitable site from which the patrol will operate. This will involve a close look at a map and other visual source material, such as aerial reconnaissance photographs. A good field of vision is of paramount importance, as is a position with a concealed route for entry and exit. The chosen location therefore must afford the 'watchers' the best possible view, while at the same time providing them with cover from the eyes of the enemy. For this reason most OPs, unless they are to be used for a short period only, are dug into the earth and sited on high ground.

The SAS team concealed themselves under a large disused van

SAS rules for choosing an OP site are quite simple. First, a position which appears as an obvious and ideal location will also appear as such to the enemy, and so is avoided. Second, the most unlikely sites may be favoured because they will arouse little interest. In Northern Ireland, for example, a hide was established in the back yard of an IRA suspect's house. The SAS team concealed themselves under a large disused van, which had been converted into a shed and used for growing mushrooms. Third, positions are never selected where the hide would look over a skyline, or where it would break an even outline.

Once a location for the patrol's OP has been selected from the map, together with at least one alternative should the first prove unsuitable for some reason, the patrol can be inserted. If the OP is to remain occupied for some time and requires solid construction, additional rations, water and stores will probably be required. In order to spread the load that will have to be man-packed into the OP area, an escort patrol might be provided, both to help carry the kit required, and to protect the OP patrol until it reaches its position. An additional benefit of being inserted as part of a larger patrol is that should the group be spotted by the enemy, it will look like a fighting patrol.

Below: An ideally-situated covert observation post allows an SAS four-man team to gather valuable intelligence concerning the strength, location and movement of enemy forces.

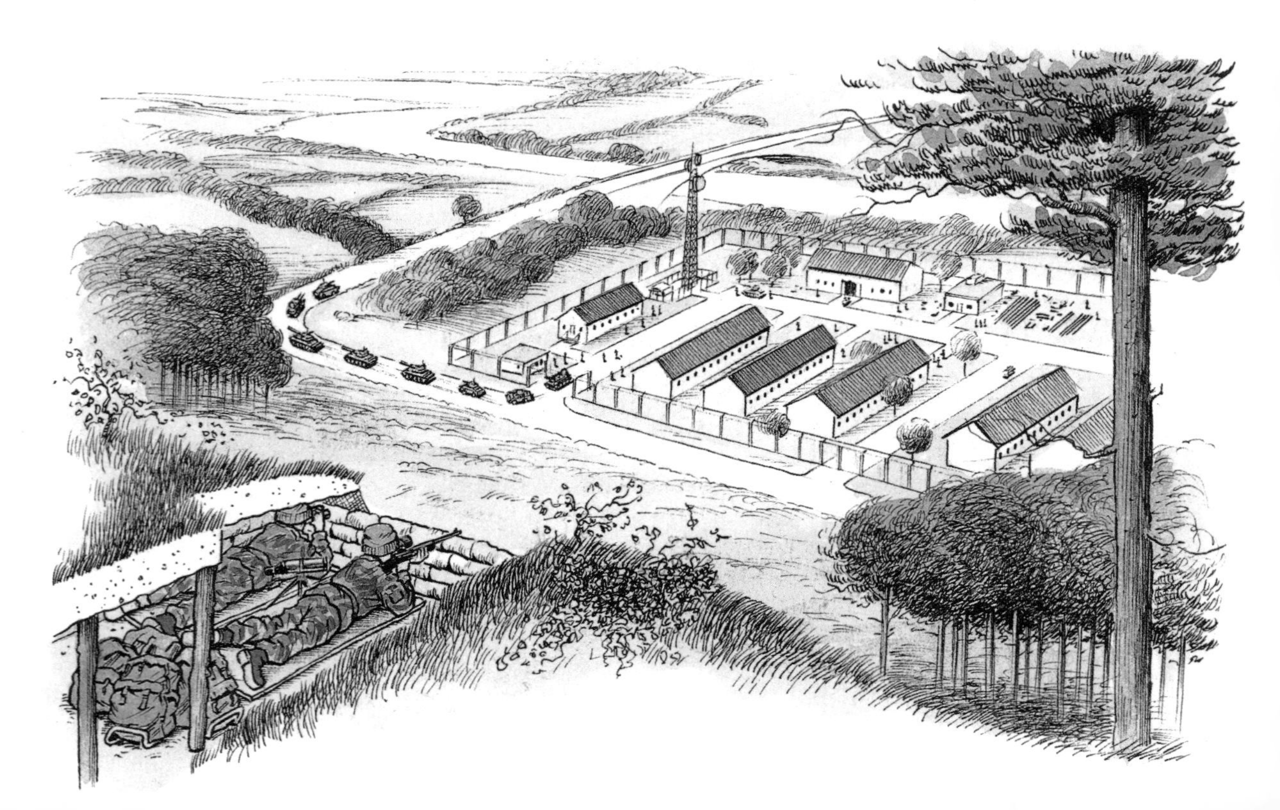

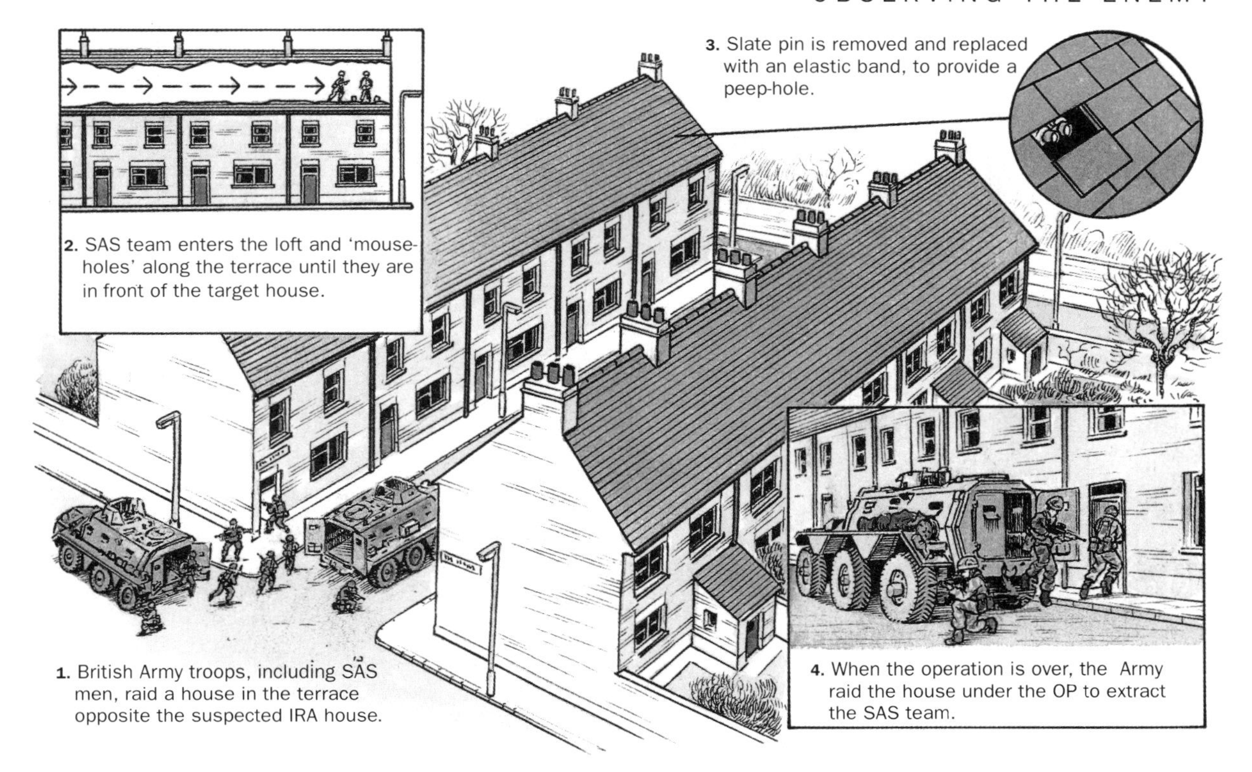

Above: Re-creation of an event which happened during an operation in Northern Ireland. OP work in Ulster is extremely tedious and dangerous.

As it is difficult to be precise about numbers at night, the fact that four men might 'go missing' will not be obvious to enemy observers.

When the four-man patrol reaches a point just short of the OP site, they will stash their bergens and adopt a defensive posture. If the OP patrol has an escort this can be left behind while the four-men move forward to conduct a recce of the actual OP position. If there is no escort, the standard procedure is for the patrol commander and lead scout to move forward and conduct the reconnaissance, leaving the remaining two men to provide security.

Once the OP has 'comms', the patrol divides into two pairs

Once the reconnaissance phase is completed the entire patrol will move into the OP position and occupy it. From the OP location the patrol signaller will establish communications with the SAS base. It is useless to have an OP in an ideal spot from which to observe the target area, if the intelligence gathered cannot be relayed back to those who require it. Once the OP has 'comms', the patrol divides into two pairs. Two troopers provide local defence while the other two begin to prepare the OP. In certain situations, such as when an escort patrol is available for some local defence, three men can dig while one remains on guard and radio watch. Short-term observation positions (24 hours or less) do not need to be dug-in, but the setting-up procedure is still far from simple. SAS camouflage and concealment techniques ensure that while the OP is being constructed, it is not detected by enemy ground or air forces. A hessian screen will be put up, together with a poncho and scrim camouflage net for overhead cover. The whole procedure is carried out at night, and emphasises the troopers' ability to co-operate closely and almost instinctively with each other, bearing in mind that speech is kept to an absolute minimum.

Exactly what materials a patrol needs to construct a covert OP depends on the operational area, but the following items are typical of an OP equipment check list: waterproof

PAIRS

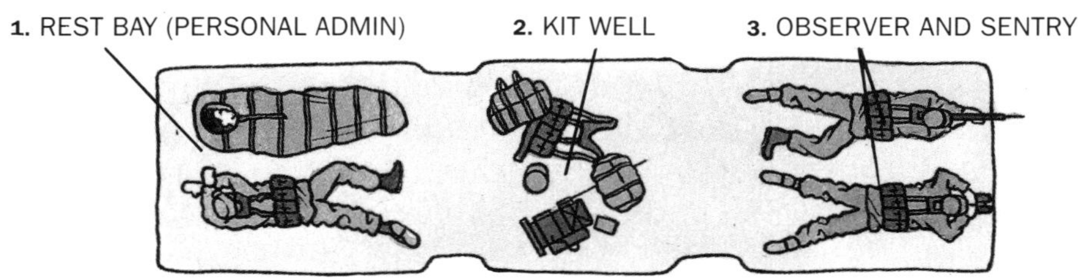

Used in linear features (fencelines and boundaries). Camouflage netting is draped over small observation ports and the observation bay is screened from any internal light source by black hessian.

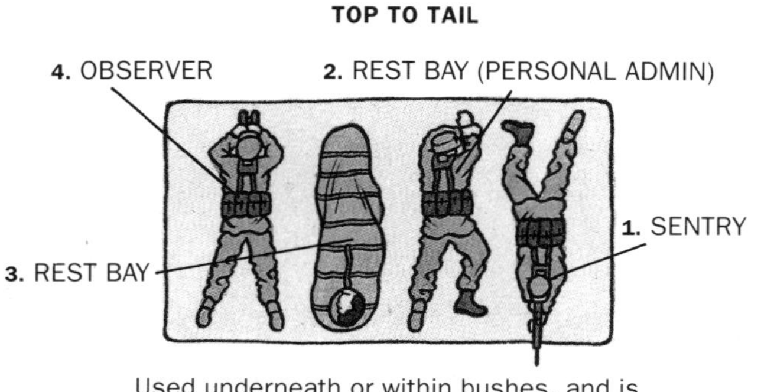

Used underneath or within bushes, and is also used along linear features. Good for all-round defence.

STAR FORMATION

1. SENTRY
5. KIT WELL
2. OBSERVER
4. REST BAY (PERSONAL ADMIN)
3. REST BAY

Type of OP depends on nature of available cover. Personnel must minimise amount of disturbance when establishing OP.

Above: The most common types of observation post layouts employed by the Regiment.

sheets or ponchos; locally-available material (doors, small logs, wooden stakes, iron pickets, chicken wire, metal sheets); cutting tools (axes, *goloks*, machetes, even secateurs); digging tools (spades and pick axes); camouflage netting and face veils; parachute cord and twine; and thick leather gloves.

There are two dug-in static OP layouts used by the SAS. The first is a simple dug-in OP in the shape of a star, with one SAS patrol member lying facing out in each 'arm'. This design is used for short-duration OPs, being semi-permanent and easy to construct, in addition to providing some overhead cover.

The simple method of construction is to dig the basic star shape with an open drainage well in the centre, into which excess water will run. The star is then covered with local material, ponchos and turf. The deeper the arms the more room the occupants will have to manoeuvre, and the larger the structure the more room for kit. The star-type shelter gives good all-round visibility, is relatively easy to construct, and is easy to 'bug-out' (escape) from. It can be occupied for as long as the OP party can put up with the confined space, and the weather.

The second type of shelter is rectangular in shape and is constructed in a similar fashion. Because it is deeper there is more overhead cover, but being larger it takes longer to dig and produces more spoil. The spoil can be removed by wrapping it in a poncho and either dragging it, or carrying it in a bergen, and dumping it out of the way some distance from the hide. Being larger it is more comfortable and can therefore be used for longer periods — perhaps a week or more. Once covered, in the same manner as the star-type and using a

Right: A more traditional British Army OP, equipped with powerful optical equipment.

combination of issue and local materials, it provides a more comfortable, concealed shelter from which the OP party can operate. Accessed by a camouflaged entry/exit hole, the larger shelter is virtually undetectable, providing it has been well-constructed and the overhead cover matches the local terrain.

For intelligence-gathering purposes the OP will be equipped with a wide range of special-to-purpose kit including optical equipment, such as binoculars, spotting telescopes, and night vision aids. They will often be covered with camouflage material or painted black, to prevent any light from reflecting off them, and thus compromising the position of the hide. The OP will also contain radio equipment, spare batteries, re-charger and codes/ciphers; log book and references (maps, aerial photos, etc.); camera and film; personal and close-support weapons; water and rations; warm clothing and sleeping bags.

Additional cover was used to hide the OP from thermal imagers

The Falklands War provided an example of the high standard of SAS observation skills. Teams were landed on the islands and constructed OPs to observe Argentinian positions prior to the British landings. One was established, for example, on Beaver Ridge, overlooking Stanley airfield. The OP was concealed with a turf roof, allowing it to blend into the surrounding ground. Additional cover was used to hide the OP from thermal imagers which Argentinian helicopter crews used during their sweeps of the hillside. The inside of the hide was always damp, despite its lining of plastic sheets, and poorly ventilated. It had only one small aperture, through which a powerful telescope provided a view of Stanley airport. The SAS men were cramped and uncomfortable, their boots were always damp and their limbs numb from constricted blood circulation. Only at night, when the risk of being spotted was at a minimum, could they safely move about and thus exercise their tired and aching limbs.

SAS OPs also played a vital role in the 1991 Gulf War against Iraq. For weeks before the Allied powers' ground attack, SAS patrols and OP parties were inserted into both Iraq and Kuwait, to identify possible targets and in some cases help neutralise them. By adjusting friendly incoming artillery fire, or guiding air-launched missiles on to enemy targets using laser designation equipment, SAS troopers manning OPs played an aggressive, as well as passive, role.

The Regiment's expert observation skills are undoubtedly finely honed. The ability of four-man patrols to establish and operate OPs involves the use of a wide range of patrol skills, concealment techniques, as well as calling upon the resilience of individual troopers. The sheer monotomy of OP work can be a severe test of individual mental stamina and determination. But another important factor is the sheer daring of patrols when choosing OP positions. In adhering to the Regiment's 'Who Dares Wins' motto, OPs are often a stunning success because they are placed in locations which, according to rational military logic, would be regarded as impossible, and are therefore overlooked by the enemy.

SIGNALS AND RADIO SKILLS

All SAS soldiers are trained to the highest standards in the art of signalling and the operation of communications equipment. In addition, the Regiment's political and military masters have ensured that they have access to the most sophisticated signalling equipment that is currently available. The kit operated by today's SAS trooper means that he can communicate with the Regiment's base at Hereford from almost any location in the world.

The importance of communications to SAS operations cannot be overstated. The collection of vital intelligence by SAS reconnaissance patrols and those involved in OP tasks would be of little value were it not made readily available to SAS commanders in the rear.

All SAS 'Sabre' Squadron troopers are trained in communications skills, and each four-man patrol has its own specialist signaller. Such is the importance of communications that all potential SAS soldiers are introduced to signalling skills during the Continuation phase of their initial training. These skills are integral to the successful

The state-of-the-art in military communications equipment — the PRC 319 radio system manufactured by Thorn EMI Electronics — as used by the SAS Regiment.

operation of the Regiment's wartime role. During the Falklands War, for example, SAS patrols, inserted on to the islands before the main landings, relayed information concerning Argentinian activity back to the Task Force's commanders aboard HMS *Hermes*. Intelligence gathered by these four-man patrols in and around the Stanley area revealed that the Argentinians were dispersing their helicopters into the hills each night, away from the airport at Stanley which was being bombed by aircraft and shelled by naval gun fire. After this information had been radioed back by the SAS, Harrier jets were able to locate the helicopters in their new locations and attack them.

All SAS troopers must be trained to Regimental Signaller standard

Radio communications allow SAS patrols to maintain contact with their operational base, pass on information and call-in artillery and air strikes, or aircraft for exfiltration once a mission is over. In modern warfare signalling has assumed an increased importance and, as a consequence, all SAS troopers must be trained to British Army Regimental Signaller standard. This involves being able to transmit and receive Morse code messages at a minimum of eight words per minute. In addition, they must master all SAS codes and ciphers, which were developed from those used by Special Operations Executive (SOE) agents operating in occupied territory during World War II. The standards demanded for continuous wave (CW) communications, codes and ciphers are higher than those required of all but the most specialised operators within the Royal Corps of Signals, the organisation which provides communications throughout the entire command system of the British Army. One indication of the SAS Regiment's professionalism in the area of military 'comms' is illustrated by the number of different systems they must be capable of operating. Most combat units, such as infantry battalions, armoured and artillery regiments, normally use a maximum of six different types of radio set. The SAS, by comparison, must be adept in the operation of some 30 various communications devices.

A special Royal Corps of Signals unit — 264 Signals Squadron — is attached to the SAS. Its personnel must pass their own selection course and qualify as military parachutists prior to joining the unit. They are all highly competent specialists who provide the SAS with its vital world-wide communications links.

Until recently the primary radio equipment used by SAS four-man patrols was a Clansman High-Frequency set, similar to that used at battalion level by British Army combat units. Often mounted in vehicles, the PRC 320 in SAS service is often man-packed by the patrol signaller. When used by ground troops it operates on compact batteries, rather than being float-charged by its carrier vehicle's electrical system. In addition, deep-penetration patrols often carry a lightweight hand-powered generator to recharge the batteries, should the need arise. Although it has a voice/speech capability, it is used in the SAS role as a Morse or CW transmitter. It has a maximum range of 100km

Below: If the antennas on their sets are damaged, SAS troopers can easily construct simple sky-wave antennas to convey their message.

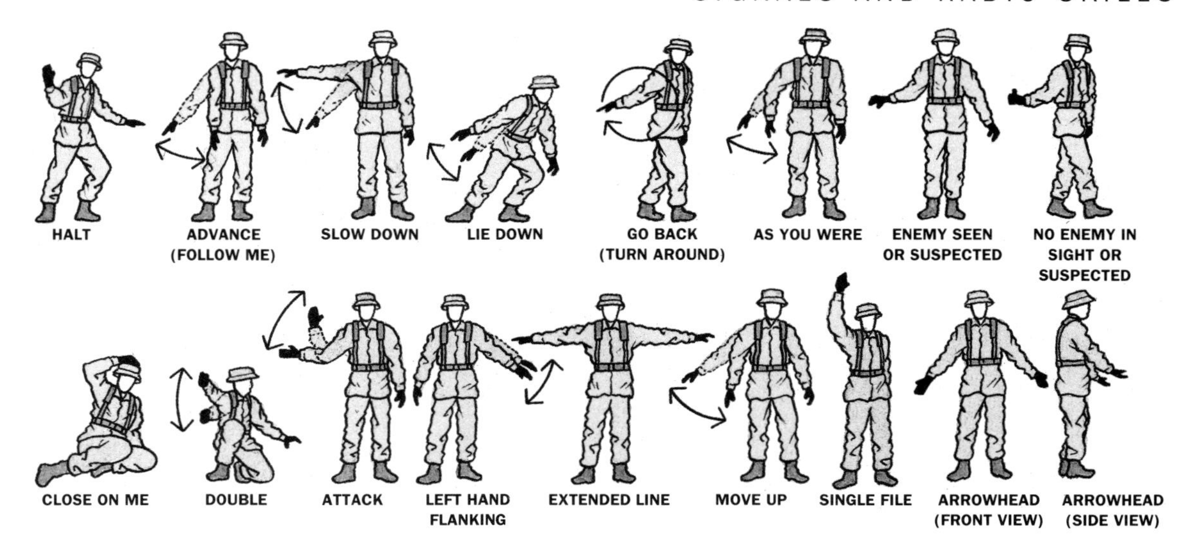

Above: Field signals, a simple and very effective way of transmitting orders throughout the patrol, also help to maintain noise discipline.

(voice) using a conventional whip antenna, but the range can become virtually limitless when operating CW in the sky-wave mode, where the operator's signal literally bounces of the earth's outer atmosphere.

Another radio set that has been gaining credibility with special forces is the PRC 319 HF/VHF set. Like the 320, the 319 is capable of sky-wave transmissions, but the latter set is lighter and more adaptable, and has been built to withstand immersion in water and being dropped by parachute. It is a powerful 50 watt tactical patrol radio transmitter/receiver with a electronic message unit, a battery, and one or two antennae tuners. The entire package measures just 205mm high, 210mm wide and 100mm deep, and weighs just 3.4kg, making it ideal for special forces (SF) operations, where weight and bulk are at a premium. Capable of data, voice or CW transmission over an extremely large frequency spectrum (1.5 to 40 MHz) over randomly selected channels via a two-way simplex system, the messages sent using this system are almost impossible for the enemy to intercept. In addition, the PRC 319 can also store up to ten pairs of channels in a protected memory, and in normal operation can send/receive via a select switch and small keypad, which has an integral liquid crystal display (LCD). A fully-automatic tuning unit (TURF) permits the use of a well-sited antenna for covert operations — allowing the signaller to send messages while located some distance from his set. Another bonus the PRC 319 offers the SAS trooper is an integral burst-message capability.

These radio systems are very robust 'soldier-proof' devices

Currently, all SAS patrol radios are normally used in conjunction with small, portable burst transmission devices, such as the MEROD manufactured by Racal of England, or the lightweight DMT built by Plessey Military Communications. These systems are very robust 'soldier-proof' devices capable of storing outgoing and incoming messages. Messages are loaded into the equipment using a keyboard, displayed and checked, then sent through a radio in a single burst by depressing one key. Available with built-in encryption facilities, such devices mean that accurate messages can be sent in one or two seconds, so fast that the sending station cannot be detected by enemy direction-finding (DF) equipment. In addition, should the message itself be recorded it will be virtually impossible to de-code.

Other recent innovations include compact man-portable ground satellite communications (SATCOM) systems, such as those produced by the specialist military electronic communica-

Above: During a hostage-rescue operation the SAS commander (top right) closely liaises with his snipers, assault groups and the civilian police.

tions company, Ferranti. Their MIL/UST-1 satellite ground terminal has been specifically developed for use by deep-penetration patrols, such as those operated by the SAS. Ideal for use in situations where normal line communications are unsuitable, such as over long distances (like those encountered by the SAS during the Falklands campaign), satellite systems have proved their worth to special forces the world over. The MIL/UST-1 comprises a hand-held ground terminal which features a microprocessor-controlled synthesiser coupled with an LCD display. As the term suggests, satellite communications systems transmit messages to their home station via an orbiting satellite, and as such are an extremely secure and effective method by which SAS patrols, troops and squadrons can transmit and receive intelligence, and other miscellaneous information. These modern systems mean that a trooper can, in theory, communicate with his headquarters at Hereford from anywhere in the world.

Despite rapid advances in radio technology in recent years, the primary method of communication between members of an SAS patrol once it has deployed into the field remains by hand signals. A well-trained and efficient four-man patrol, especially one that has operated together as a team for some time, has little need to vocalise. It works together almost instinctively, each patrol member knowing exactly what is expected of him in almost any given situation, and doing it without being asked to do so.

The standard SAS method is 'hands-on' using a Morse key

Most instructions and information required to be passed within the patrol can be done through a series of silent hand signals, most of which need only one hand to be used, allowing the individual signalling to maintain a firm grip on his personal weapon. Signals to control patrol movement, its order of march and to identify possible threats, are known to all members of a patrol and are practised prior to each mission during the preparation phase.

If verbal instructions are needed, or if for any reason members of the patrol have to converse, it is usually done in low, hushed tones, which do not carry as far as higher-pitched whispers. The only exception to these standard operating procedures (SOP) concerning vocal communication is in the event of a contact with the enemy, in which case verbal orders are shouted so they can be heard over the noise of gunfire. Indeed, while communications within the patrol is done mostly by hand, so too is its communications with the outside world. Rather than use voice when sending messages back to base by radio, the standard SAS method is 'hands-on' using a Morse key. The code has been used by Britain's SAS as its primary method of communication since the Regiment's early days in the desert of north Africa. Morse has a greater range than voice transmissions, and it can be encoded, making it an ideal means by which to pass secret messages. The code is, above all, secure.

There is a need for good communications in the area of CT warfare

Quite apart from the Regiment's requirement for communications in its more 'conventional' operations, such as deep-penetration patrols and behind-the-lines reconnaissance missions, there is a need for good communications in the area of counter-terrorist (CT) warfare, specifically, during hostage-rescue operations. Two distinct departments, or 'cells', within the SAS Group are responsible for the technical battle to counter terrorism: the Operational Research Wing (ORW) and the Counter Revolutionary Warfare (CRW) Wing. The ORW cell, formed in the early 1960s, has influenced the design of many specialist technologies, from waterproof fabrics and textiles, to sophisticated electronic communications equipment.

The need for advanced personal communications systems between CRW assault and back-up teams, and their controllers, led to the development of new, powerful two-way radio sets capable of being used within buildings while wearing respirators, and operated without the need to take hands off weapons. The exact nature of such equipment, its operating frequencies and design remains, like so many items of SAS 'sneaky kit', a closely-guarded secret known only to those in the CRW teams and the ORW cell.

The assault entry team will also use signals to coordinate their movements

However, the deployment of this equipment during a hostage-rescue operation is known. In any anti-terrorist operation the SAS commander on the ground will quickly establish a communications network to allow him to control every aspect of the these highly volatile situations. Initially, he will establish telephone communications between his on-site command post, any tactical command posts, and the overall operations centre. If, for any reason, a telephone network cannot be established, then communication will usually be by field phones or radios. Throughout the entire operation communications will also be maintained with the relevant civil and military authorities.

The commander will then establish call signs and codes, which will be relayed to all SAS troopers, whether they are part of the perimeter containment group or members of the assault groups. The communications network, thus established, ensures simultaneous relaying of information from the command centre to snipers, entry elements and vice versa. Before, and during, the assault entry team members will use also hand and arm signals to coordinate their movements.

Communications are an integral part of SAS operations, be they long-range reconnaissance patrols, OP work or counter-terrorist missions. The great increases in the range and sophistication of communications equipment means that SAS missions can be more tightly controlled, thus ensuring a higher degree of tactical flexibility, especially during anti-terrorist operations. In addition, expert skill in the use of many different types of signals equipment ensures that the SAS are second to none in this particular field.

BEHIND ENEMY LINES

During the 1991 Gulf War SAS patrols, working closely with their American counterparts, fulfilled numerous vital tasks behind Iraqi lines — assisting Allied aircraft in their strikes, destroying communications lines and helping in the rescue of downed friendly aircrew. The special forces' various missions made a valuable contribution to the eventual Allied victory. For the SAS it was a return to their World War II role, and one more example of their expertise at fighting deep behind enemy lines.

In war the majority of soldiers fight with friendly units on their flanks and behind them, and with the enemy in front of them. SAS troopers, on the other hand, are trained to fight a different kind of war. It is a military maxim that small, highly mobile forces operating behind enemy lines can inflict damage out of all proportion to their size. The SAS was devised by David Stirling for just such a role, and this task remains integral to the Regiment's operational doctrine. In a conventional conflict, such as the one fought in the Persian Gulf in 1991 to liberate Kuwait, SAS teams are inserted behind enemy lines to cause disruption

Troops operating behind enemy lines require constant supplies of food, ammunition and medicines. Here, a Blackburn Beverley aircraft drops provisions to SAS patrols in Malaya.

through raids, sabotage and ambushes, to gather intelligence, and to organise resistance groups.

These teams would strike at the enemy's relatively ill-defended rear areas. Targets would include airfields, headquarters, field repair workshops, communications centres, vehicle parks and supply dumps. Such targets offer themselves as easy prey for a unit prepared to take the risks involved in reaching them. In World War II, for example, most Axis war material arrived in north Africa by ship before being sent to supply depots, from where it could be transported to the front. Together with headquarters and air bases, such supply centres made ideal targets for those who dared attack them. In November 1941 the SAS, operating behind the enemy's front line, carried out four weeks of raiding on the north African coast road used to transport Axis supplies, destroying large quantities of trucks and their cargoes. Between 6 June and 6 September 1944, SAS patrols operating deep behind German lines in central France cut railway lines, derailed six trains and ambushed a large number of supply columns. Other SAS raiding forces in France hit bridges, telephone exchanges and electricity sub-stations, often conducting their raids with the aid of local resistance fighters, the *Maquis*, or Special Operations Executive (SOE) agents.

Very little information escapes the attention of the 'Kremlin'

Operations mounted by the SAS require a very high level of prior planning and preparation. The SAS Group's operational intelligence (OPS/INT) cell is responsible for the collation of intelligence and its dissemination to SAS planners. This information usually includes details on every aspect of the prospective theatre of operations: local geography, climate, political infrastructure, government, economy, police and armed forces, etc. In short, OPS/INT can provide all the facts necessary to plan, prepare and, if necessary, fight almost any type of campaign anywhere in the world.

The intelligence itself is gleaned from a number of sources including numerous British, American and West European intelligence sources (which include satellite and aerial reconnaissance photography); the press; radio monitoring stations (such as GCHQ); and any political opposition groups and dissident figures from the country in question. In the area of Counter Revolutionary Warfare (CRW) operations, the SAS liaise very closely with a number of outside agencies including NATO and friendly non-aligned intelligence services, and numerous counter-terrorist groups such as Italy's NOCS, Germany's GSG 9 and America's CIA and FBI. Very little information escapes the attention of the 'Kremlin', as the OPS/INT cell is affectionately known, whether is concerns a terrorist group or a Third World country. The need to know the enemy is important not only to senior commanders, but also to individual SAS troopers as well. They are the soldiers who actually conduct the operations, and within the SAS Regiment there is a tradition of its members, regardless of rank, knowing exactly what they are doing, and why.

The SAS have the trained men and the technology required to pull it off

The SAS organisational system, with four specialist troops within each of the four squadrons, gives its commanders a high degree of flexibility when planning an active mission. They can select soldiers with experience in at least one Troop specialisation: Mobility, Boat, Mountain or High-Altitude/Low-Opening (HALO) parachute operations. Whether the mission is a raid, an ambush, a reconnaissance or any other type of operation which demands unconventional forces, the SAS have the trained men, the expert techniques, and the technology required to pull it off.

The basic SAS procedure for operating behind enemy lines has changed little since World War II. Once a target has been identified a plan of action will be formulated. A force will then be inserted behind enemy lines to infiltrate the target area. Finally, if an assault is required, the target will be attacked using either surprise, coupled with intense speed and firepower, or by sneaking up undetected and planting delayed explosive devices before retiring unseen. Both these methods require a great

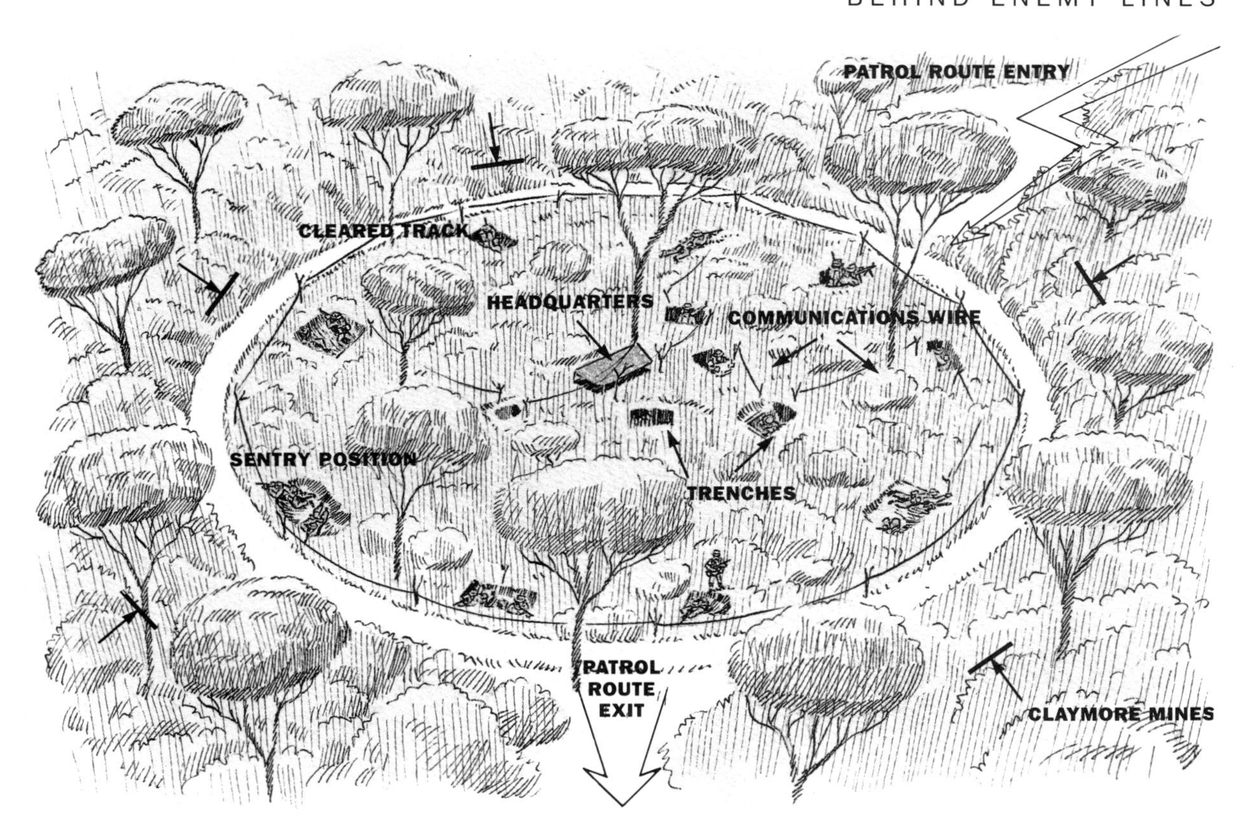

Above: An SAS jungle base camp protected by Claymore mines and sentry posts.

deal of courage, ingenuity, resourcefulness and outright skill.

Once behind enemy lines the ground rules for SAS patrols are relatively simple, but because of the risks involved they place a great strain upon the soldiers taking part. Being on constant alert for any sign of enemy forces, endeavouring to move as silently as possible without leaving any tell-tale signs, and often weighed down with heavy loads, is mentally and physically exhausting. Such a situation can be a supreme test of an individual trooper's stamina and strength. During an SAS assault upon an enemy-held airfield in Oman in 1970, for example, each patrol member, moving through rocky terrain on foot in high temperatures, carried 400 rounds of GPMG ammunition wrapped around his body, 600 more in his bergen, as well as four SLR magazines for his personal weapon, each holding 20 rounds. In all, each trooper was carrying 40kg of weapons and ammunition, in addition to three water bottles , rations and personal kit.

Most movement is conducted at night to avoid detection by enemy aerial reconnaissance aircraft or helicopters. When travelling through hilly terrain, SAS teams, either on foot or in vehicles, will keep clear of ridges to avoid being silhouetted on the skyline. Bridges, roads and tracks are also avoided if possible, as they are often patrolled, mined or booby trapped. Maps carried by patrol members will not be marked or folded in such a way as to show the target area. If a soldier is captured with this information it will compromise the whole mission (care is also taken not to mark the map with soiled fingers, as this will highlight the target as clearly as a pencil).

Today, the SAS use the Lockheed Hercules heavy transport aircraft

After insertion, the first requirement for an SAS 'Sabre' Squadron is to establish a secure base from which offensive operations can be

Above: A World War II two-man SAS jeep equipped with twin Vickers 'K' machine guns.

mounted. This process is normally started by an advance party which either begins to take over from the previous squadron, or selects a suitable site to establish a new one. There are normally two types of base: a headquarters base area, which is often co-located with a major strategic terminal such as a port or airfield: and a forward operational base (FOB), normally sited on or near a tactical landing strip, or at least a good landing site (LS). Headquarters base areas and FOBs have always been used by the SAS in its post-World War II campaigns, the exception being during the early stages of the Falklands conflict. In this war the headquarters base started off on Ascension Island, before being transferred aboard ship with the British Task Force operating in the South Atlantic. FOBs are usually well-defended, dug-in and have overhead cover. In addition, their construction allows defence with a fraction of the occupying force while the remainder is on patrol, an important factor if most of the base is away on operations for most of the time.

All additional spares, weapons, ammunition and other equipment for the mission are usually airlifted to SAS forces operating behind enemy lines, as and when required. During the Malayan campaign Blackburn Beverley aircraft were used to drop supplies to SAS teams operating deep in the jungle. With a range of 2080km and a payload capacity of 20,500kg, the Beverley was an extremely capable aircraft. Today the SAS use the Lockheed C-130 Hercules heavy transport aircraft; its payload is double that of the Beverley and it also has a longer range. Dropping supplies by parachute can have its drawbacks: pallets can drift for kilometres in strong winds and can be lost. In addition, loads, if not packed properly, can be damaged. During the 1970 campaign in Oman, SAS teams operating in the mountains were re-supplied with ammunition by airdrops. Some containers carrying mortar bombs, which were badly packed, fell on to the rocks, resulting in them exploding and injuring one trooper.

How they will actually achieve their objective is left entirely to the SAS forces on the ground. For instance, D Squadron, deployed during the Falklands campaign, had to attain certain objectives, most of which were reconnaissance in nature. Working directly with the senior commanders of Britain's Task Force, the SAS squadron commander selected the area, or the specific target, and allocated it to a troop. The troop commander then selected one or two patrols to carry out the particular mission. Once the patrol was on its own, the responsibility for exactly what action to take rested solely with the patrol commander (PC). He could be advised of factors affecting the general situation by those back at base, or offered specific-to-mission advice by members of his own patrol. Ultimately, however, he was in the best position to judge how to approach the target, what to look for and what measures to take. The PC may only be a corporal but he can still be the man in control.

Larger patrols are normally commanded by troop leaders

Larger patrols, such as those which mounted cross-border operations into Indonesia during the violent Borneo confrontation, are normally commanded by troop leaders (usually captains) or, in some cases, squadron commanders. However, the principle remained the same: he was the 'boss' on the ground once the patrol

Below: The modern SAS Land Rover. The launchers at the front can fire flares and smoke grenades.

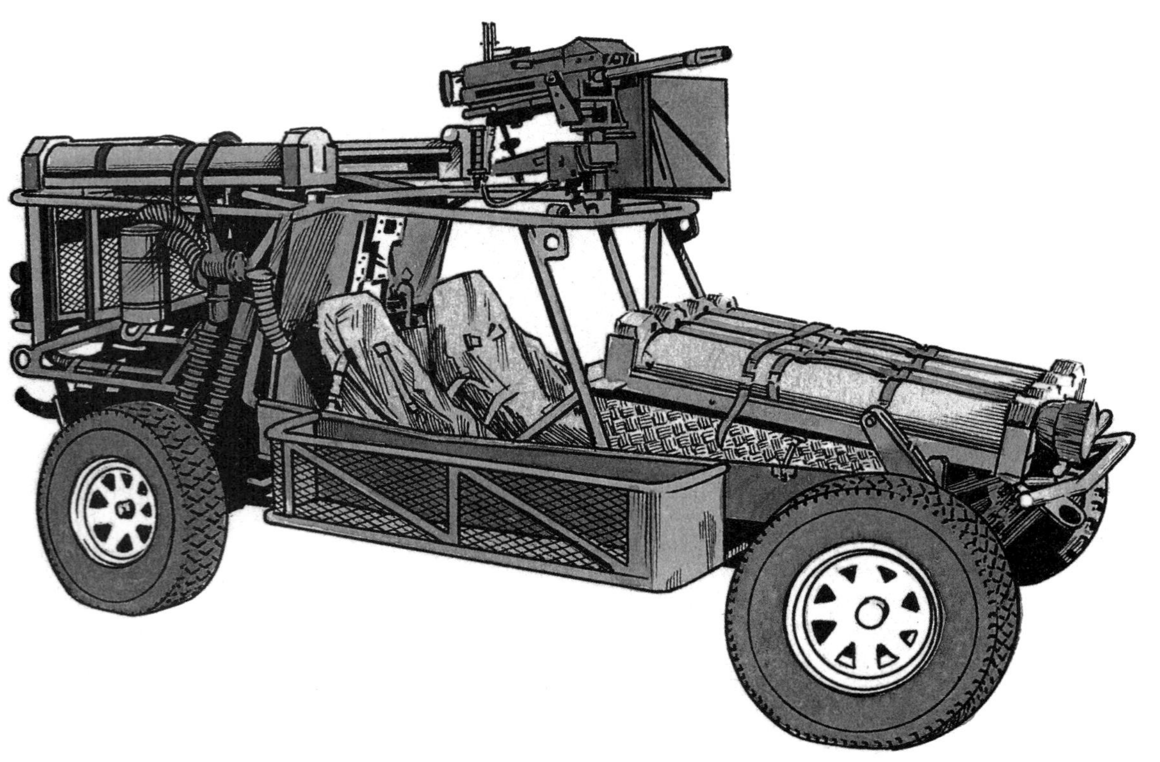

Above: An Iraqi aircraft shelter destroyed by a laser-guided bomb. Such precision bombing was often carried out with the assistance of the SAS.

had deployed. Nevertheless, as far as both internal counter-insurgency and cross-border operations are concerned, neither can be mounted with a good chance of success without proper planning, effective command and control, or a well-established base camp.

One SAS patrol spent 103 consecutive days in the hostile jungle environment

Once the SAS has established a firm base from which to operate, missions can be mounted. Together with long-range reconnaissance, the most common SAS mission behind enemy lines is the deep-penetration fighting patrol, a type of operation in which Britain's SAS has had plenty of practical experience. In Malaya, for example, one SAS patrol spent 103 consecutive days in the hostile jungle environment, its only contact with the outside world being by radio communications.

Above left: Re-creation of an SAS night-time attack on an Iraqi airfield during the Gulf War. Left: The Light Strike Vehicle.

Another type of operation in which Britain's SAS Regiment reigns supreme is border control and monitoring. Enemy guerrilla forces can receive supplies or reinforcements across border areas, especially if the border is long and runs through inhospitable terrain, and they can also withdraw to the safety of a 'neutral' country when required. Control over border areas, therefore, needs to be established if the battle against hostile insurgents is to be won. Locally-raised frontier guards may, if properly supported, be able to tackle this task and secure the border. If the threat is too great, these local guards may be tasked with acting as guides for more regular units.

During the Regiment's activities in Malaya there was little opportunity to secure the border areas and the communist terrorists (CTs), as the guerrillas were known, had to be isolated and attacked in-country. In Borneo, as

1. Movement to the objective.

3. Withdrawal after raid. One security element draws off pursuing enemy forces while the other covers the withdrawal.

2. Action in the objective area.

Above: Diagrammatic representation of how a troop-sized unit would undertake a raid on a strategically-important target behind enemy lines.

the confrontation escalated, SAS units of up to squadron strength were authorised to cross the border in so-called 'Claret' operations, and strike the enemy in locations where they had previously considered themselves safe from attack. In the period January to March 1965, for example, patrols from B and D Squadrons mounted successful cross-border missions. In these particular cases the 'insurgents' were in fact regular Indonesian troops, for by this time the guerrilla groups that had been actually involved in the earlier stages of the campaign had been successfully dealt with inside Borneo.

The success of SAS operations behind enemy lines can often depend upon the ability to move rapidly from one area to another, both to avoid enemy forces and to hit targets in quick succession. Therefore, ever since the first SAS raids in December 1941, the Regiment has used a wide variety of special vehicles for mobile operations. In World War II SAS teams used jeeps to attack enemy targets. The ubiquitous jeep was well suited to Special Air Service operations, being fast, highly mobile and very manoeuvrable. They were designed to carry a crew of two, but by the end of the war had been substantially modified for fighting behind German lines, as Sergeant Bob Bennett, one of the original SAS members, describes: 'The whole front of each jeep was covered in armour plate....Armament consisted of twin Vickers [machine guns] for the front gunner, another pair for the rear gunner and every third or fourth jeep carried a .50 Browning heavy machine gun. The driver had a Bren gun.' Capable of putting-down an incredibly heavy weight of fire in a short period, these jeeps wreaked havoc upon enemy supply columns, troop concentrations, headquarters and the like, causing severe disruption in the rear areas. They also tied up large numbers of troops in guard and search operations, thereby denying their employment at the front.

Modern transport for operations behind enemy lines includes a wide variety of special purpose Land Rover vehicles, modified for long-range operations. These vehicles have now replaced the venerable Land Rover 'Pink Panthers', so-called because they were painted in matt pink camouflage for desert missions. In addition, in early 1991 the Regiment began to operate a new dune buggy-shaped vehicle called the Light Strike Vehicle (LSV), produced by the British company Longline. Designed to be a mobile weapons platform for long-range patrols, it has a two-man crew and was used recently in the Gulf War, when they operated on reconnaissance missions inside Iraq.

Patrol and troop skills are a major factor in the success of any mission

The LSV has a number of significant advantages for special forces (SF) work: it is easy to conceal, it can use use small tracks and access points (barriers, passes, gates and bridges), and it can also be manhandled by its crew if it gets stuck in the mud. In addition, its low silhouette and open-frame construction reduces radar and thermal signatures. The LSV's small size means it is nimble and so presents a harder target to hit than a larger, slower vehicle. The 500-600kg payload is split between food and water, ammunition and other stores. Ammunition alone accounts for 300kg (2000 0.5in rounds, 350 30mm rounds, 400 40mm grenades or 5000 7.62mm rounds). The LSV can mount Milan or TOW anti-tank weapons, with carrying capacity for between six and ten reloads. Other weapons which can be mounted on the vehicle include a 40mm grenade launcher, a 30mm cannon and a 0.50 calibre gatling gun. In addition, allowance has been made for overloads — other troops, casualties and equipment — as was experienced during recent operations in the Gulf (on one mission an LSV carried 12 men with their personal kit, in addition to its normal load). The LSV's top speed of 100km/hr allows it to hit hard and then withdraw before the enemy has chance to retaliate.

Patrol and troop skills are a major factor in the ultimate success of any mission an SAS unit may undertake but, as with reconnaissance patrols, the secret behind the success of a fighting patrol depends largely on planning and preparation, organisation, and delegation. Fighting patrols involve more men — 16 or more — than a four-man reconnaissance/observation post (OP) patrol, and so teamwork and proper co-ordination are especially important. The mission sequence is also more complex. After a briefing and kit checks, the fighting patrol will be divided into separate groups, each with its own individual area of responsibility for the forthcoming operation. Most fighting patrols are divided into three groups: the assault group, the fire support group, and the forward rendezvous (FRV) party.

The assault group is the largest, comprising the bulk of the patrol and would normally be relatively lightly armed, perhaps with just personal weapons and grenades. The second largest section would be the fire support group, probably armed with one of more GPMGs or the equivalent and M203 grenade launchers. The third, and smallest, group would be the FRV party which might, in some cases, operate one or more mortars to provide additional support for the other two groups. The role of the assault group would be to attack the target. The fire support group would put down covering fire to assist the assault group and cover its withdrawal, while the FRV party would secure a point on line of immediate withdrawal from the target area.

The role of the assault group would be to attack the target

A classic example of a successful SAS fighting patrol action took place in 1982 during Operation 'Corporate', the code-name for Britain's campaign to retake the Falklands. Up until 14/15 May 1982, less than six weeks after the Argentine invasion of the islands, the SAS had confined itself mostly to passive missions, mainly fielding reconnaissance patrols to gather intelligence information. On the night of 14/15 May the emphasis changed, with the SAS conducting a major raid upon the Argentine air base on Pebble Island off the northeast coast of West Falkland. The airfield was the base for

Pucara ground-attack aircraft, each having the capacity to carry a very large quantity of aerial ordnance, and capable of causing a great deal of destruction among the British troops who were to land on the islands six days later.

The unit given the task of 'taking out' the target was D Squadron 22 SAS. Members of D Squadron's Boat Troop conducted the initial reconnaissance of the airstrip itself, after being inserted on to Pebble Island by canoe and establishing an OP overlooking the Argentinian base, which contained 11 aircraft plus a number of buildings for personnel and stores.

M203 grenades and 66mm LAW rounds struck the stationary aircraft

While plans for the attack were being finalised the OP party established a mortar base-plate and an FRV, and settled in to await the arrival of a remainder of their squadron. This force was flown in by three Royal Navy Sea King helicopters on the evening of 14 May, and landed at an LS some 10km from the airstrip. All three aircraft arrived safety and unobserved after a 45-minute flight from HMS *Hermes* and, after a very brief on-the-spot assessment, the troopers marched towards the FRV. The six kilometres cross-country 'tab' passed without any incident, and on arrival at the mortar base-plate each trooper dropped off the two 81mm mortar bombs he had been carrying in addition to his personal weapon and ammunition. The troopers then continued their march a further four kilometres to the FRV, where they divided into individual troops and were guided to their starting-off positions by members of the Boat Troop's reconnaissance party. So far everything had gone according to plan.

The actual assault on the airfield was spearheaded by Mountain Troop, which opened fire on reaching the enemy base's perimeter. The surprise was total. The entire area was illuminated by para-flares fired by the naval support vessel, HMS *Glamorgan*, and the 81mm mortar. Simultaneously, M203 grenades and 66mm LAW rounds struck the stationary aircraft. Those that were left relatively undamaged by D Squadron's first devastating barrage were soon destroyed by explosive charges placed on them by individual troopers, or were engaged at close-range with small arms and anti-tank fire.

With all the targets destroyed the SAS began an orderly withdrawal under cover of mortar and naval artillery fire. The Argentines attempted one counterattack during the withdrawal phase, but were beaten back after the officer leading them was cut down by a well-aimed burst of machine gun fire. SAS casualties were incredibly light: one trooper with shrapnel wounds and another slightly injured by a small command-detonated landmine. The raid had been a resounding success, with all enemy aircraft destroyed and a large number of their personnel killed and wounded. The runway had been cratered, making it useless for other fixed-wing aircraft, and a large quantity of military ordnance had been destroyed.

Above: A Sikorsky HH-3 helicopter receives fuel from a Lockheed HC-130 aircraft. Such in-flight refuelling of helicopters means that special forces can be inserted deep inside enemy territory.

The 1991 Gulf War between Iraq and the forces of the United Nations again illustrated the diversity of missions capable of being undertaken by SAS units behind enemy lines. Two 'Sabre' Squadrons, half of the Regiment's fighting strength, were deployed to operate in occupied Kuwait and deep inside Iraq. Roving around the vast desert areas and infiltrating urban areas, often disguised as Arabs and speaking Arabic, SAS teams carried out such missions as intelligence gathering, assisting conventional forces and launching many hit-and-run attacks against Iraqi military targets.

To help American, British and French aircraft make accurate strikes against targets located in Iraqi urban areas, SAS two- and four-man teams infiltrated the cities of Baghdad, Basra and Kuwait. These teams were equipped with small, portable, battery-powered laser designators, which resemble unfurled umbrellas in appearance. Once they had found the target, after moving through the streets disguised as Arabs, the SAS would direct a designator at it from a concealed location nearby. Once it was turned on the troopers would leave the area. The designator would emit an invisible laser beam, which would hit the target and bounce off into the sky. The 'smart' bombs carried by Allied aircraft would then use the beam to home in on the target. In this way the SAS assisted in the destruction of strategically-

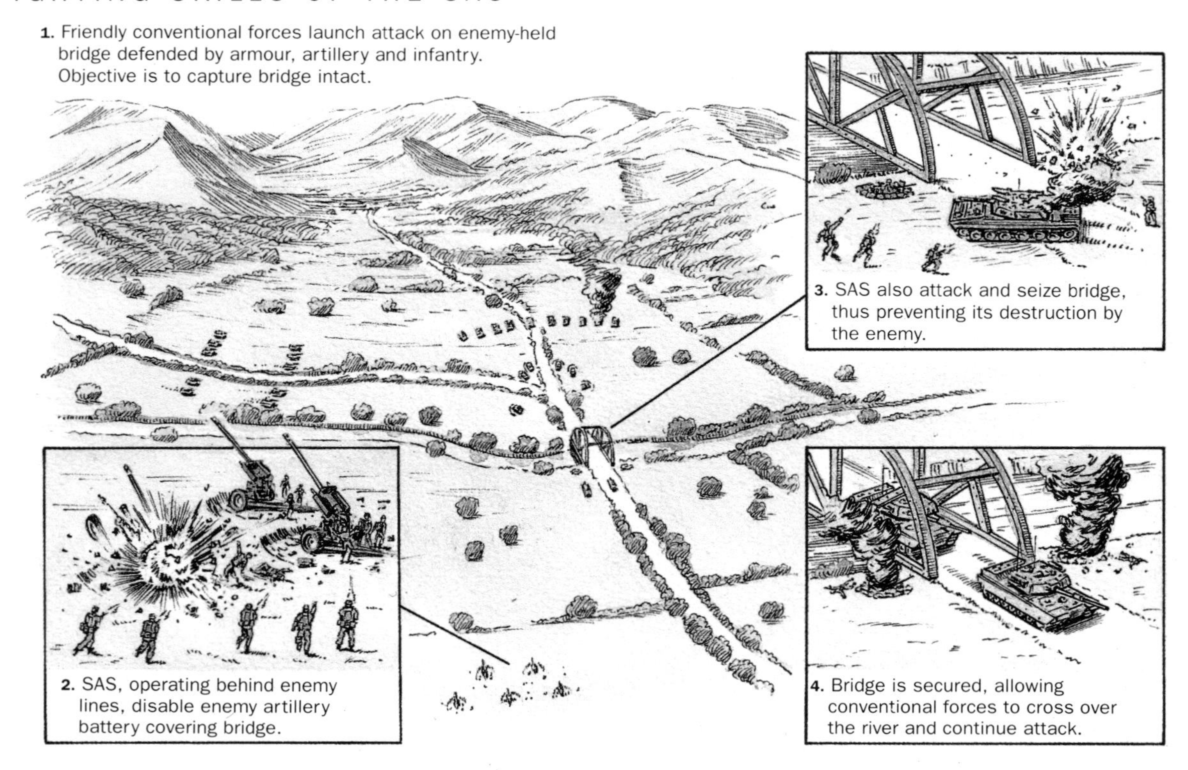

Above: Diagrammatic representation of how SAS teams operating behind enemy lines can assist conventional forces in seizing a strategic target.

important bunkers and other command centres.

In the desert of western Iraq the SAS, operating in LSVs and Land Rovers, hunted down mobile Iraqi Scud surface-to-surface missile (SSM) launchers. Large numbers of these missiles, which were being fired at Israel in an effort to drag the Jewish state into the war, and their launch vehicles were destroyed, either by aircraft assisted by the aforementioned designators, or by anti-armour weapons mounted on LSVs. The ability of the LSV to move rapidly through the desert also resulted in the SAS being used to search for downed British and American pilots, who were often found before Iraqi troops reached them and then evacuated by helicopters called in by the LSV's radio.

SAS patrols also kidnapped several Iraqi officers

Prior to the Allied land assault, the SAS played an important part in locating Iraqi communication lines linking Kuwait to a command centre in Basra. The SAS succeeded in finding them and then blowing them up, forcing the Iraqis to use radio signals. These were then picked up by the SAS and the intelligence collected was then given to Allied HQ in the Saudi capital, Riyadh.

The SAS were also heavily involved in OP work, constructing camouflaged hides near Republican Guard and regular Iraqi Army units, thereby being able to provide accurate intelligence on the whereabouts of supposedly concealed enemy units. Other tasks involved stealing enemy surface-to-air missile (SAM) systems and even a Soviet-built Mig-23 Flogger interceptor aircraft, both of which were required for threat-assessment purposes. SAS patrols also kidnapped several Iraqi officers who were then shipped to Saudi Arabia for interrogation. Other officers, known to be political supporters of Saddam Hussein, were assassinated to lower the morale of front line units. The latter missions were referred to by General Schwarzkopf, the American Commander, Middle East, when he confirmed that 'direct action missions' had been undertaken by SF units (American SF teams were also conducting similar operations throughout the war).

One traditional SAS tactic while operating behind enemy lines has been liaison with friendly partisan forces. During the Gulf War, for example, SAS teams would have formed contacts with the Kuwaiti Freedom Fighters for intelligence gathering purposes. In 1944 the SAS, operating in France, were frequently in close land cooperation with the *Maquis*. In the summer of that year they had established a joint camp at Fontainbleau. It was then betrayed by a member of the *Maquis* and attacked by the Germans; only after sustained savage fighting did the SAS manage to escape. As a consequence of episodes such as this all troopers are taught to exhibit caution when dealing with partisans, and it is now usual practice not to share camps with such forces. The usefulness of partisan forces will always be weighed against their general unreliability, ill-discipline and disorganisation.

The SAS will do their best to win over the local population to their cause, as this can reap substantial military rewards. This is the 'hearts and minds' policy which was so successful in Malaya, Borneo and Oman. The tribespeople along the border of Borneo and Indonesia, for example, were remotely located, and SAS four-man patrols lived in and operated from their hamlets for extended periods, often for four months at a time. The subsequent trust formed between the tribesmen and the SAS, mainly as a result of the medic's work, reaped many rewards. Together with the natives, the 'local' SAS troopers gathered useful information about enemy movements and, when the situation warranted it, joined forces with other SAS units and struck the Indonesian insurgents in raids and ambushes on both sides of the border.

'Captured special forces troops must be handed over to the Gestapo'

SF soldiers, including the SAS, run a high risk of being either captured or killed when operating behind enemy lines. Their sabotage and ambush operations make them extremely unpopular with enemy commanders, and this can result in them being ill-treated, or even

Below: Representation of how forward operational bases are used behind enemy lines.

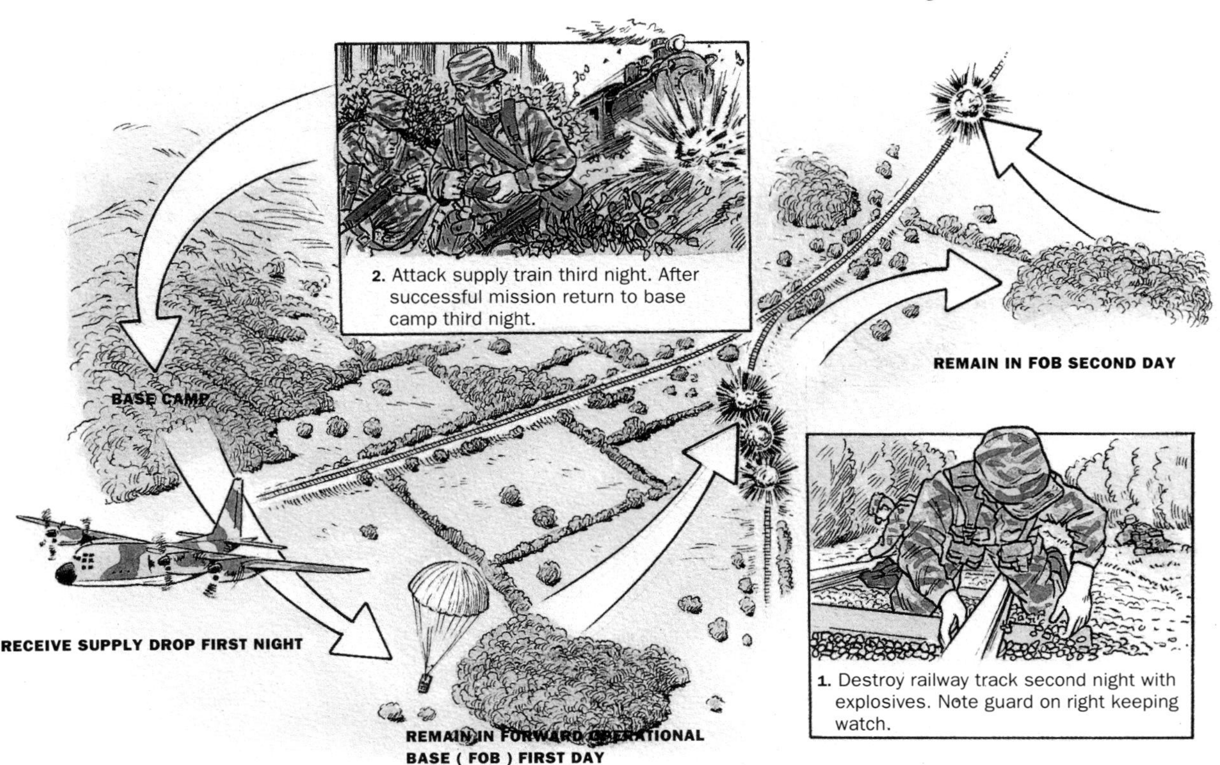

killed, if captured. Their status as prisoners of war is very dubious, especially if they are captured wearing disguise or civilian clothing. During World War II the German supreme commander, Adolf Hitler, issued a personal directive to his theatre commanders stating that 'captured special forces troops must be handed over at once to the nearest Gestapo unit.....these men are very dangerous, and the presence of special forces troops in any area must be immediately reported.....they must be ruthlessly exterminated.'

SAS soldiers will usually be interrogated before they are incarcerated

This directive, initially issued in response to the many SAS successes in the north African desert, but not implemented by Rommel during his time as commander of Germany's Afrika Korps, singled out SAS soldiers as different, and directed that they should be treated accordingly. Not all German Army commanders were as humane as Rommel, and later in the war SAS personnel captured in occupied Europe could expect little mercy. Many were treated in accordance to Hitler's directive, and were tortured and shot after capture.

If captured SAS soldiers will usually be interrogated before they are incarcerated or killed, to extract any information concerning the strength of friendly units and mission objectives. Therefore, at Stirling Lines all troopers are instructed in how to resist interrogation. Adhering to a few simple rules can mean the difference between being broken down or successfully resisting the interrogators. First, upon capture all troopers are instructed not to resist being tied and bound, this will only waste precious energy which can be used to resist the subsequent interrogation. Second, they are taught to give the impression they are more tired than they actually are. There is a

A group of South Vietnamese soldiers interrogate a terrified suspect during the Vietnam War. Modern mind-bending drugs have resulted in techniques much more effective than the crude water-torture method being administered here, making resistance much more difficult.

simple reason for this: during questioning no interrogator wants an unconscious prisoner on his hands; if he thinks the captive will collapse then he will temper the interview accordingly. This will often mean no further beatings or shouting, at best it could result in the interrogation being terminated, all be it only for a short while.

The Geneva Convention requires that a captured soldier gives his name, serial number and date of birth, but invariably the interviewer will want more. This is where the energy saved earlier will come in useful, for the interviewer will employ a host of sensory-deprivation and psychological methods to extract information. In such a situation the trooper's mind must be functioning clearly.

In some cases the enemy will have access to mind-bending drugs

If the prisoner does not divulge any information he will often be physically abused. This can mean being spread-eagled against a wall and made to stand on the toes and fingertips, which causes severe pain and disorientation. The trooper will start to look forward to sitting down and facing a barrage of questions, rather than being tortured in this way. This means he is more amenable to divulging vital intelligence. It is impressed on individual troopers not to give in, but rather to call upon their reserves of strength and mental determination.

In some cases the enemy will have access to mind-bending drugs which can break the strongest of wills. Consequently, every effort is made to mount operations to rescue the prisoners before they divulge valuable information.

However, SAS expertise in concealment, stealth-like movement and general fieldcraft when operating in enemy territory, means that very few SAS troopers are taken prisoner. Even when patrols are cornered their superior weapons skills, discipline and coolness under fire often ensures they can hold off the enemy long enough to escape. The fact that less than a dozen SAS men were killed or captured during the 1991 Gulf War is testimony to the supreme skill displayed by men of the Special Air Service when fighting behind enemy lines.

STAYING HIDDEN

The ability to remain undetected while on active service is an important part of SAS operational procedure. Great emphasis is placed on learning the methods by which troopers can successfully camouflage both themselves and their equipment. For small units working on their own in hostile territory, the ability to remain hidden is imperative to their success. In many different types of terrain the Regiment has proved itself adept in the art of concealment.

The trooper's world was starting to disintegrate. A member of G Squadron 22 SAS, he had been landed by Royal Navy Sea King helicopter some days previously. The place was East Falkland and it was early May 1982. Following the Argentine invasion and occupation of the British dependencies, special forces (SF) patrols had been inserted prior to the main British landings to the northeast of the major island of the group: East Falkland. The trooper now in big trouble was a member of such a patrol.

Tasked with locating numerous enemy positions, defences, strengths and dispositions, assessing their fighting ability and morale, SAS reconnaissance patrols moved stealthily around the island, unobserved by the occupying

Get to know the ground you are operating over, adapt your dress accordingly, and carefully select your routes when moving around to maximise your concealment.

Argentine forces. Often in appalling weather, conditions most patrols were dropped two or more days march from their final target area, and moved at night after resting during the day. The type of lying-up position (LUP) used by these patrols were mainly individual shell-scrapes, small hollows covered with wire, on top of which was an overlay of local vegetation.

The vegetation forming the roof of his hide started to shred away in the gale

The soldier in question had become aware of an Argentine helicopter cruising the area a few minutes before. Lying face down and unable to see what was happening directly above him, the man on the ground could only remain motionless while the increased down-draft, caused by the aircraft's main rotor at such low level, began to reveal his position. The vegetation forming the roof of his hide started to shred away in the gale, leaving him increasingly exposed with little more than his SAS wind-proof smock and trousers for camouflage. Once the helicopter had moved off the soldier quickly set about concealing his presence once more. Within a few minutes there was nothing on the open, seemingly empty, ridge line to suggest to the Argentinians that their enemy had landed.

The success of many SAS missions depends to a large extent upon the ability to remain concealed from the enemy. Stealth and secrecy allow the SAS presence and intentions to remain hidden, and unknown, by their foe until it is too late. However, in order to achieve the surprise needed to overcome the odds, much depends on the individual SAS soldier's field-craft skills, among which are those of camouflage and concealment.

There are three main principles employed in the concealment of personnel, objects and military activity: siting, discipline and construction. Without the proper use of all three concealment measures undertaken by military personnel, be they regular British Army or SAS, will be to no avail.

Siting includes the selection of the most advantageous position in which to hide a man, an object or an activity. Even most apparently featureless terrain contains some pattern which can be used to conceal signs of military activity. However, siting is governed by three factors. First, mission consideration. A site may be perfect from the concealment point of view, but if it makes it impossible to undertake the mission, it is pointless. Second, the requirement for dispersion dictates the size of the site. A site which does not allow enough dispersion for effective operation is useless. Third, there are considerations of terrain patterns. The position must not disturb the terrain patterns in which it is sited; this is imperative for the purpose of deceiving enemy reconnaissance aircraft.

Strict camouflage discipline must be maintained by the unit as a whole at all times, as well as individual soldiers. Carelessness and laxness will have the inevitable result of revealing the position to the enemy. New tracks should always follow existing paths, roads, fences, or natural lines in the terrain pattern. In addition, there must be strict observation of blackout procedures. Camouflage discipline must be maintained at night as well as during the day.

Defensive positions will be laid out in depth around the perimeter

The third principle, camouflage construction, includes the use of artificial or natural materials to help blend the object or individual with the surrounding terrain. Camouflage construction should only be resorted to when siting and discipline cannot produce the optimum concealment required. As well as blending into the surroundings, the materials used must be able to withstand varying weather conditions.

SAS units, like all British combat troops, receive intensive instruction in how best to construct defensive positions for between two to four men, and how to make the best possible use of all available firepower from a well-arranged position covered by at least two arcs of fire. Though such positions are primarily designed for ground troops, such as deployed infantry, rather than smaller SF units, the basic principles of camouflage and concealment

Right: Proper concealment depends on the correct use of surrounding terrain. Some examples of the right and wrong positions.

WRONG
RIGHT

WRONG
RIGHT

WRONG
RIGHT

RIGHT
WRONG

RIGHT
WRONG

WRONG
RIGHT

RIGHT
WRONG

WRONG
RIGHT

Above: New tracks (left) indicate activity; using an old track (right) arouses no suspicion.

remain the same.

In counter-insurgency (COIN) operations, units of up to squadron strength may construct a forward operational base (FOB) which, because it will most likely be co-located with an airstrip or helicopter landing site (LS), will be difficult to conceal from the enemy. In such cases, defensive positions will be laid out in depth around the perimeter, to slow down and absorb any attack. Individual positions will also be concealed and located in such a manner as to provide mutual support. They will also be constructed to enable the base to be defended by a third of its compliment. This is important as SAS tactics involve deploying the majority of the force outside such bases. This means fewer men are available to defend an SAS FOB from an external threat than might normally be expected for a base of this size.

On the move an SAS patrol relies on stealth to remain undetected

Individual defensive positions must be concealed from observation from both the air and the ground. To be effective they must be well camouflaged and well sited. Careful appreciation and proper use of the ground results in any attacking/advancing enemy coming within effective range of the position while remaining completely unaware of its existence. The resulting action is usually short, violent and inflicts heavy casualties upon the enemy. One of the most suitable sites for a defensive position, for example, is on a reverse slope, just behind the crest of a hill, so that direct enemy fire can only be brought to bear once the enemy has crossed the skyline and presented the defender with an ideal target. These rules apply regardless of whether the defended position is a part of an FOB or a troop/patrol LUP.

For a patrol LUP, the need for concealment is even more important. Their restricted firepower means that all other available assets have to be used to their best advantage. On the move an SAS patrol relies on stealth to remain undetected, but once stationary it seeks shelter as soon as possible. The patrol will choose a site which allows effective use of personal weapons, as well as offering protection from observation and enemy attack. LUPs normally rely on personal camouflage and local vegetation to effectively conceal their occupants. In temperate climates hedgerows and gorse thickets are among those locations suitable for LUPs, as they provide good cover and are not obvious hiding places.

To blend in naturally with his local environment an SAS trooper must have very highly-developed personal camouflage and concealment skills. These skills are just as vital to surviving on the battlefield as weapons skills. The aim of combat survival is to remain undetected from the oldest, most primitive battlefield sensor: the human eye. Far more common than thermal imagers, passive night goggles (PNGs), infra-red sights and a host of other hi-tech devices, the human eye still poses the greatest threat to the SAS trooper on active duty.

In order to disguise himself, the soldier needs to remember the basic fieldcraft rules of concealment and camouflage: shape, shine, silhouette, sound, smell, colour and association. The first principle is to disguise the shape, both of the man himself and the tools of his trade. Helmet, bergen, webbing, weapons, radios and most other military kit tend to have an obvious shape — often square with sharp sides or straight lines — and stand out against the surrounding 'natural' environment. By attaching local material, such as foliage, leaves or grasses, to webbing and other equipment a trooper can blend into his surroundings, without looking like a 'walking bush'.

For most troops the helmet is the first item of equipment to be camouflaged, by inserting local vegetation into the elasticated straps sown into the disruptive pattern material (DPM) cover which breaks up the distinctive outline of the helmet. However, the SAS soldier rarely, if ever, wears a helmet, except when parachuting.

Below: Structures made from natural materials, such as this jungle basha, blend into the background and are often easy to build.

Headgear worn on missions depends on the type of terrain in which a unit is operating. Bush hats are worn in the jungle, and cloth, peaked combat hats in temperate and other climates. The beige SAS beret is never worn in the field, it being reserved for barracks and headquarters duties when applicable.

Cloth strips can also be used to break up the outline of the barrel

To disguise the shape of his equipment and to merge into the background, the SAS trooper can use various artificial and natural aids. Neutral, green or brown cloth, or hessian, can be attached to loops on webbing or netting. Waterproofed DPM camouflaged covers are available and can be used to break the outline of bergens. Camouflaged nets, normally used on vehicles, can be cut down to size and employed to cover personal hides. In situations where there is little local vegetation, commercial chicken wire can be used and, once bent into shape, can form a semi-rigid 'roof' on which peat or turf can be laid.

The rifle is one of the most distinctive features carried by the modern fighting soldier, and therefore the SAS trooper will camouflage his personal weapon. The easiest way to do this is with paint; SAS personnel operating in the Falklands could often be identified returning from patrol by the green-painted M16s they were carrying. During the 1991 Gulf War

1. Construction of the basha. Branches cut from surrounding trees can be used to make a simple framework. Large jungle leaves are used as covering material.

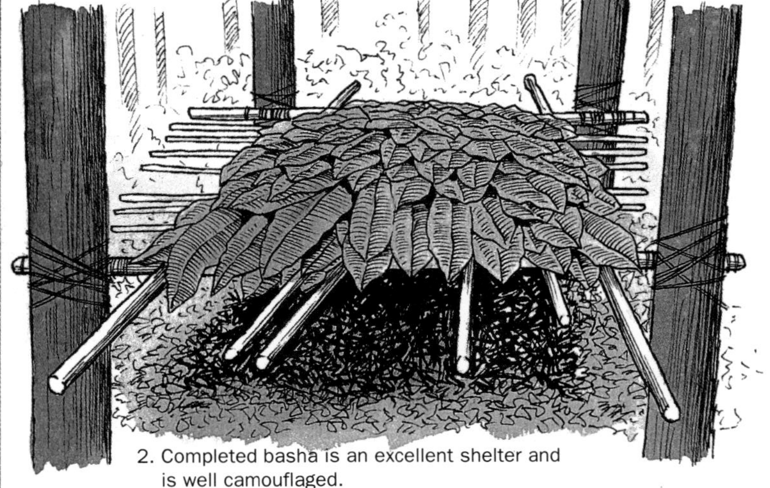

2. Completed basha is an excellent shelter and is well camouflaged.

Allied SF units and others, including certain British infantry battalions, adopted the practice of painting their personal weapons a sand colour.

An effective way of distorting the outline of a weapon is by wrapping it with strips of cloth dyed to match the background, though care must be taken to ensure that nothing interferes with the weapon's working parts and sights. Cloth strips can also be used to break up the outline of the barrel while the magazine, pistol grip and top cover can be taped. In temperate and tropical climates, additional camouflage in the form of masking or 'jungle' tape, together with DPM cloth, can be used to make a man's personal weapon even less visible at extremely close ranges.

Below: The pattern on this tropical combat dress corresponds to the shadows formed in the jungle.

These measures are also effective at reducing shine, one of the most common reasons for troops being spotted by an observant enemy. Until the 1950s British Army soldiers wore brass on their uniform, both on the parade ground and in battle. Brass buckles, used to hold and fix webbing together, were, on the insistence of some unit commanders, kept clean and shiny. Such surfaces acted as an obvious giveaway when on active duty. Luckily, the 'bullshit' factor has never been a part of SAS routine, and today the rest of the Army has followed its example. In the field, at least, all such surfaces are parkerised, chemically dulled or painted. However, despite the use of plastics and coated metals, the SAS soldier still has to guard against shiny surfaces on the modern battlefield.

Skin catches and reflects light and therefore requires camouflage

Optics such as weapon sights, binoculars and spectacles are all 'signature equipment'. Even the surfaces of prismatic compasses reflect the light and can reveal a patrol's position. Some equipment, such as mess tins and eating utensils, can be dulled by applying paint at regular intervals. The lenses on optical equipment, on the other hand, cannot be tampered with and must be used with care. Binoculars, for instance, can be used from behind a face veil, allowing almost 100 per cent vision while almost totally cutting out shine, and when not in use can be stowed inside the owner's camouflage smock.

One of the least obvious objects that reflects light, even at night when the moon is visible, is the human face. Skin, of any colour, catches and reflects light and therefore requires camouflage. Various techniques were developed during the early stages of World War II, starting with Allied commandos who 'blackened-up' their skin to reduce shine. Burnt cork was a popular method, being applied to the face and hands to cut down reflection and disrupt the overall shape. Most modern armies use a cosmetic camouflage ('cam') cream to cover exposed areas of skin. 'Cam' cream is designed to cut through the face's vertical and horizontal features — eyes, nose and mouth — making them less obvious and more difficult to identify even at very close ranges.

Below: Stick camouflage is used to cover the exposed parts of the body — the face, neck and hands — and is extremely effective.

Facial camouflage comes in various colours. It is normally applied in three stages. First, a thin base coating, diluted with water or saliva, is used to cover all exposed skin areas. A common mistake is to miss backs of hands, ears and neck, but patrol members usually pair off to check each other's personal camouflage as part of their normal operational routine. Once the face and hands are dulled, the next stage is to

break up the shape and outline of the features. Any format will suffice, the most simple being a diagonal pattern across the face. The final phase of applying facial camouflage is to darken normally high-lighted areas — forehead, nose, cheek bones and chin — while leaving areas normally in shadow a lighter shade. A mesh mosquito netting can also be used to break up the outlines of the face.

Washing of the body and clothing helps minimise obvious odours

Another important part of concealment, not always obvious but nevertheless important, concerns identifying and disguising smells. Although the olfactory gland is perhaps the least used of human senses, it is surprising how much a man's sense of smell improves after a few days in the field. Body odour may be the first thing that is noticed, especially if personal hygiene has been minimal, either through lack of discipline in the case of poorly-trained conscript troops, or circumstance, which might be the case for an SAS patrol after several days in a confined OP. Washing of the body and clothing helps minimise obvious odours, but scented soap products should not be used, and deodorants are definitely out. With regard to the latter, it has been suggested that during the early stages of America's campaign in Vietnam some US troops were detected by the VC purely on the grounds of scent alone.

Other smells are much harder to minimise. Cooking odours, in particular, are often difficult to disguise. Wood smoke, petroleum-based fuel and certain types of food can carry a great distance, depending on prevailing wind and weather conditions. Use of underground stoves, smokeless and odourless fuels, and confining cooking to the hours of daylight when the air is warmer and other stronger smells abound, helps minimise the risk of detection.

While smoke can be seen unless well hidden, smoking cigarettes can also produces dangerous by-products, and the refuse of both

Right: Weapons can be effectively camouflaged by using either cloth (a), mud (b) or tape (c). Below right: Good use of camouflage netting.

is easily distinguished by their unique smells. Burnt litter and other rubbish should be containerised and removed from the immediate operational area; it should only be buried as a last resort. If discovered it could give the enemy a clue as to the patrol's identity, as well as its size and composition. The same applies to other equipment, such as dead radio batteries. SAS standard operating procedures (SOPs) demand strict discipline with regard to refuse removal and the rule is: 'take it with you'.

Sound is another obvious giveaway. During Continuation Training all potential troopers are made aware of the measures required to reduce noise to the minimum. The 'buddy-

Right: A British Army soldier in standard disruptive pattern material (DPM) uniform. Unlike standard SAS practice, he is wearing a helmet.

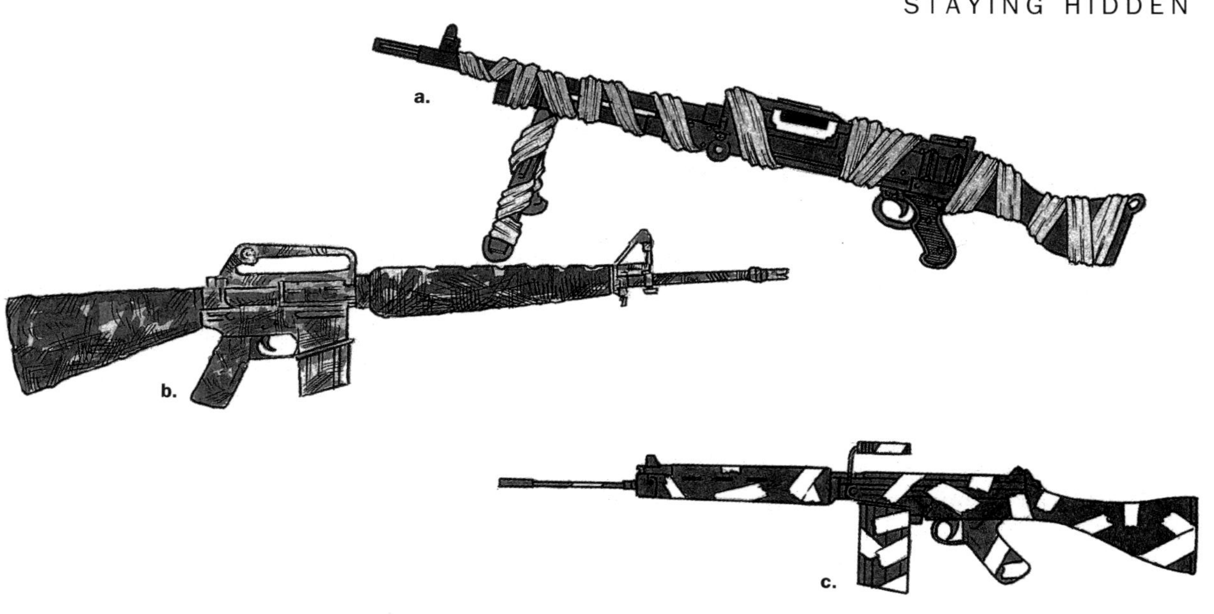
a.
b.
c.

buddy' system, whereby men pair off to look after each other, ensures that troopers check their comrade's kit, personal camouflage, etc., and have their own checked in turn, before every patrol. Boots and heavily-loaded webbing can creak, magazines can rattle, clothing can rustle and weapons can clink. Proper packing of kit can reduce noise to a minimum and 'jungle' tape can, if correctly applied, prevent metal surfaces from grating against each other. Sling swivels, which can make a noise when patrolling, are replaced with para-cord.

Fewer men make less noise and leave little trace of their occupation

Other sounds, such as radio static or 'mush', can disclose the position of the most stealthy patrol or well-concealed OP party. Difficult to disguise, the noise levels can nevertheless be reduced by proper radio procedures and discipline. The fact that the SAS makes extensive use of continuous wave (CW) or Morse, transmitted through a key and received through a headset, makes their radio communications virtually silent.

Radios also prove a problem in relation to another aspect of concealment, that of association. While a very high standard of personal camouflage among troopers will help patrols avoid detection, obvious accessories such as radio antennas can disclose the position of troops to an observant enemy. SAS patrols have a distinct advantage over most other small section-sized units as far as antennas are concerned, in that they use diapoles, or horizontal antennas, rather than vertical aerials used by conventional units. A mass of vertical radio aerials is an obvious giveaway to the precise location of a military communications centre, a fact often used to advantage by aggressive SAS patrols.

One of the advantages of the SAS four-man patrol is that it is physically easy to conceal. Fewer men make less noise and leave little trace of their occupation of a position, or their passing. In fact, the concealment of tracks plays an important part in general patrol discipline and SOPs. Tracking and counter-tracking are essential skills of an SF trooper, and many members of the SAS mastered these skills in the jungles of Malaya and Borneo during the early post-World War II campaigns.

Above: The above method of stacking and covering boxes of supplies is an effective method of concealing them from enemy aircraft.

The SAS continues to train soldiers in the skills of tracking during the jungle phase of Continuation. There are five points to successful tracking: slow and steady movement, so as not to overlook any signs; the use of the senses of smell, hearing and sight; interpreting the signs, not just seeing and following them; understanding the methods and SOPs of the enemy; and being patient and persistent. Tracking demands a high degree of concentration and common sense, coupled with careful observation. Knowing what to look for when tracking can also be of immense benefit to the patrol trying to avoid being tracked themselves.

The most common signs created by troops travelling across temperate or tropical terrain include disturbed grass and other vegetation, damaged spiders' webs, soil or clay prints left behind on hard surfaces, scuffed or scarred vines or tree bark, and disturbed ground litter and overturned stones, both of which are recognised by their darker, damp undersides being turned up. Concealing such tell-tale signs can mean a big difference when trying to stay one jump ahead of an alert enemy. The patrol's 'Tailend Charlie' ensures that there are no obvious tell-tale signs of the patrol's passing, such as upside-down leaves or bent and broken fern stems.

The colour of vehicles can be changed by applying mud to the bodywork

SAS units using Land Rovers, Light Strike Vehicles (LSVs) and other means of transport must be skilled in vehicle camouflage techniques. This is especially important if the team is not to be spotted by enemy reconnaissance aircraft. When parked all vehicles should be under natural cover if possible. If this is not possible then they should be parked so that their shapes disappear into the surroundings. Other measures include using natural materials, such as foliage, to break up the shadows and shapes of the vehicles. The colour of the vehicles can be changed by applying mud to the bodywork. Leaves and grass can also be attached to their surfaces by adhesive tape, though as they can wither rapidly care must be taken to continually inspect them and, if necessary, replace them with fresh vegetation.

Pattern painting can be a valuable supplement to other camouflage measures. Patterns are designed to disrupt the cube shape of the vehicle from all angles of view. White or light grey paint, when applied to the under surface, causes them to reflect light, thus lightening the dark shadows of the undercarriage. Vehicles can also be covered with drape nets, which are easy to use and quickly erected and taken down. They are not recommended for snow areas, however, as they require excessive maintenance, cannot support a snow load and, if wet, can become frozen and bulky, and thus very difficult to handle.

Successful camouflage and concealment is not just a case of wearing a DPM smock and trousers. In order to remain invisible and hidden from their enemies, SAS soldiers must be capable of exhibiting a wide range of camouflage and concealment skills. These include the careful selection of positions and routes, the camouflage and construction of shelters and hides, with the preference of natural materials over artificial materials; disguising or stopping 'emissions' such as smells, smoke, vapour and sound; concealing vehicles; and strict personal discipline and awareness. Successful concealment will give an SAS unit the element of surprise should it wish to launch an attack upon the enemy, or to remain hidden should the aim of its mission be passive. Failure to maintain such skills could perhaps lead to a patrol being 'compromised' and sustaining casualties, something the SAS try to avoid.

Above: Foliage can be used in concealment, such as to cover the open end of this cargo truck.

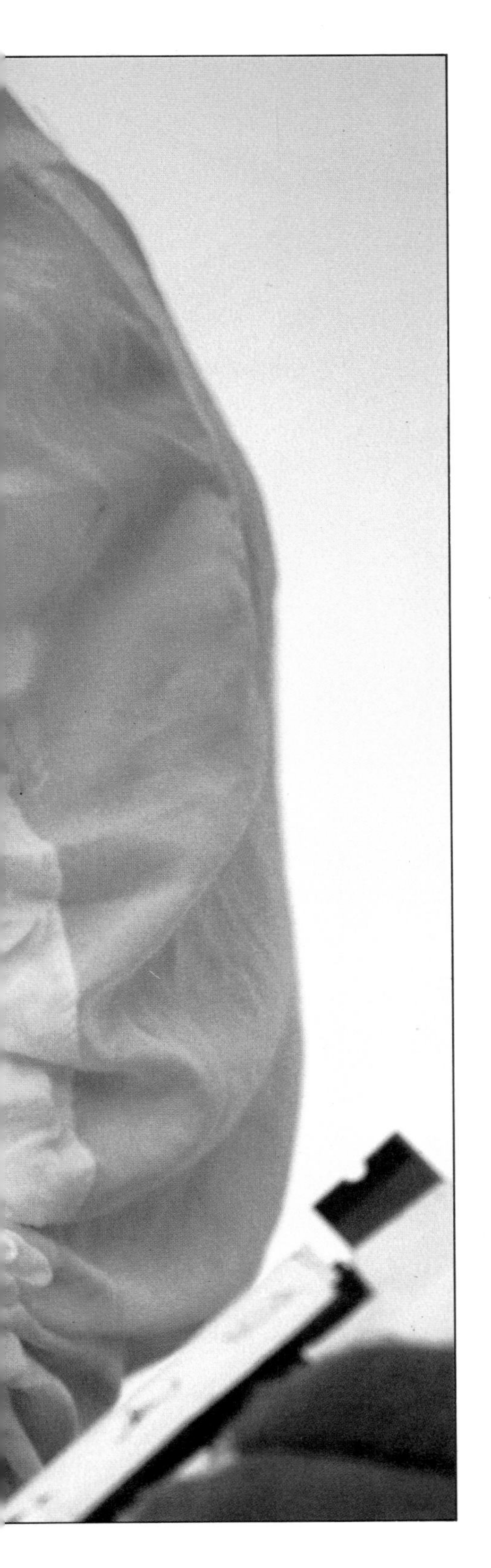

OPERATING IN HOSTILE REGIONS

One of the great tactical strengths of the Special Air Service Regiment is its ability to mount military operations in some of the world's most savage environments. Modern, high-tech equipment, together with the resilience and expert training of individual troopers, gives its commanders an elite unit that can fight anywhere, irrespective of the type of terrain encountered.

All the Regiment's soldiers undergo intensive training in hostile environments. The operational requirement for SAS soldiers to be able to undertake a mission in any environment, often at a moment's notice, means that they must be experts in fighting and surviving in many different types of terrain. The cross-training given to individual SAS troopers ensures that the Regiment is ready to operate in any theatre of the world.

One indication of the complex diversity of SAS training, and operational commitments, is illustrated by the events immediately prior to its involvement in the Falklands campaign in 1982. When the war started one of the two squadrons eventually deployed was just completing arctic warfare training in Norway, while a large proportion of the other had just begun training in the jungle. The fact

An SAS soldier, face and hands protected against the severe arctic weather, prepares to undertake a mission.The Regiment ensures that its men are equipped with the very best kit.

that those SAS soldiers who were landed on South Georgia's Fortuna Glacier were arctic and mountain warfare specialists, helped them survive the devastating sub-zero blizzard that raged through their first night, and which caused two of the three Royal Navy helicopters sent to recover them to crash before any extraction could take place.

MOUNTAIN AREAS

Mountain regions are characterised by high winds, inclement weather, difficult terrain and almost no food sources, shelter or water. Members of D Squadron's Mountain Troop encountered typically severe conditions when they were landed on Fortuna Glacier on 21 April 1982 during the Falklands War. They were buffeted by 80km/hr winds which stung the eyes of the men not wearing goggles, and

Right: Rappelling by means of a rope allows rapid descent down most vertical surfaces. During the descent the rope is passed through the left hand. Below: A picture that amply conveys the hostile nature of mountainous regions. SAS soldiers on patrol in the Radfan region of Aden in 1965.

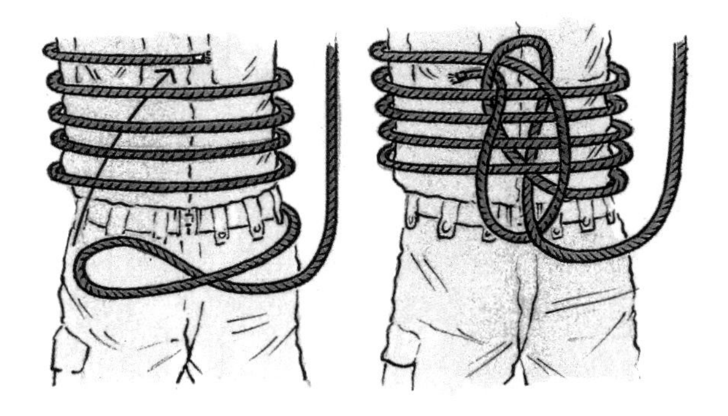

Right: The safety bowline on a coil is used by the end soldier on a climbing rope to take up the extra and unnecessary slack.

also choked the mechanisms of their GPMGs. They had to be extracted the next day by helicopter, as hurricane force 11 winds threatened them with hypothermia and frostbite.

To operate in mountain areas SAS soldiers must be skilled in many specialist techniques including rock climbing, rappelling, belaying and obstacle negotiation. Movement through mountainous terrain can be perilous. Rocks and fallen logs, for example, if kicked loose can kill men below, and can also make a great deal of noise, thus compromising the team's security. All troopers therefore receive instruction in the skills of mountain walking, whereby they learn the need to step over, rather than on top of, obstacles. This not only reduces the risk of disturbing loose rocks and debris, it also lessens the likelihood of troopers injuring themselves. A man with even a slightly sprained ankle is severely handicapped in mountainous terrain, and will be a liability on a mission.

Supplies dropped by parachute can be blown off course by high winds

Warm, lightweight windproof clothing; tough footwear and webbing; sturdy large-capacity bergen rucksacks; and high-calorie rations are all important items for conducting mountain warfare. Much of the standard-issue SAS clothing and equipment has been designed with operations in mountainous regions in mind.

The difficulties of inserting and extracting patrols operating in mountainous regions and in getting supplies to them once they have been deployed are all major problems that mountain warfare troops and their commanders must overcome to succeed in their mission. Because of the lack of good roads, for example, patrols have to be re-supplied by aircraft. Helicopters, with their ability to hover for long periods, are more useful than aircraft, as there are often very few suitable sites for landing strips. Supplies dropped by parachute can often be blown off course by high winds, or be smashed on the rocks when they land.

DESERT REGIONS

Deserts present different obstacles to SAS operations than mountainous areas, though no less difficult. Extremes of temperatures, ranging from freezing cold at night to a stifling heat during the day, are typical conditions encountered in a desert environment. Although the classic concept of desert terrain is one of sand dunes or flat expanses of featureless ground, broken occasionally by low rocky ridges and isolated hillocks, it does in fact encompass a wider range of geographical features, including mountains and rocky plateaus. Iraq, for example, which covers an area of some 435,000 square kilometres, comprises 75 per cent desert, 13 per cent arable land, 9 per cent meadow and only 3 per cent forest, while Kuwait is almost entirely desert. The area of ground operations during the Gulf War of 1991 was largely restricted to low, featureless desert varying from 91m above sea level in the north to 274m above sea level in the south, with the height decreasing towards the east and the Gulf. The road heading northeast out of Kuwait City, used by retreating Iraqi forces in February, is an undulating gravel plain (*dibdibba*) with broad wadis, long shallow depressions, small hills, and ridges. Except for the Kuwait City itself, the area is classic desert. In many of its campaigns — north Africa, the Radfan and Iraq — the SAS Regiment has demonstrated its ability to operate in all types of desert environment.

Operations in the desert are physically demanding. Dehydration, sunstroke, sunburn and the absence of local water supplies mean that SAS soldiers must adopt a number of procedures and skills, both to remain alive and to fight. During the campaign in Oman in

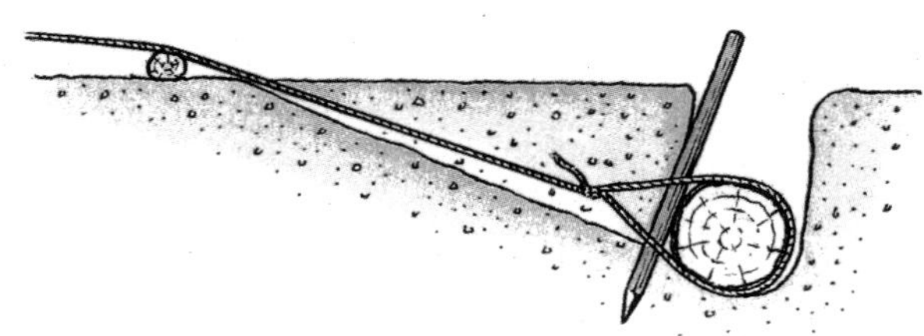

Above: Two examples of artificial rope anchors used during rappelling: the single timber deadman (right) and the picket holdfast (left).

1959, the heat was so intense that several British soldiers, all non-SAS, died of heat exhaustion while on operations. Among the basic skills required in the desert are strict water discipline, thorough personal hygiene and a high standard of physical fitness and stamina. While these qualities are inherent in the archetypal SAS soldier, the first two are especially important in hot, dry climates. Drinking and eating the correct quantities is especially important in desert operations as thirst and appetite are adversely effected by heat. It is important to drink before feeling thirsty as intense thirst is an indication that the body has already started to dehydrate. It is not unusual for a trooper to require up to 14 litres of water a day. In addition to plenty of water, extra salt must be added to food to replace that lost through evaporation. Alcohol, as well as dulling the senses, aids dehydration and increases susceptibility to heat.

Lice, mites and flies can carry diseases such as scrub typhus and dysentery

Other dangers to the individual trooper come from the various types of animals that inhabit desert regions, especially snakes — cobras, vipers, rattlesnakes and adders — which can administer lethal bites. In addition, invertebrates are drawn to man as a source of moisture or food. Spiders and scorpions can inflict painful, though rarely fatal, stings, and lice, mites and flies can carry diseases such as scrub typhus and dysentery.

Proper clothing is essential when operating in the desert. During the day it must protect the trooper from the sun, at night it must keep him warm, and at all times it must be a barrier against insect bites and thorns. While some standard kit may be adapted, much is specifically designed for use in desert conditions. Clothing should be light and loose fitting to aid the prevention of water loss, while attention to head and footwear is particularly important. Members of Britain's SAS Regiment have long regarded the traditional Arab headdress as the most ideal form of headgear for desert operations. The *shemagh* is no more than a fringed, cotton wrap-around scarf which is used to protect the eyes, nose, mouth and neck from sun and sand (eyes can also be protected from the intense sunlight and sand by tinted goggles).

Sanitation is particularly important for SAS patrols operating in the desert, as intestinal diseases are an ever-present threat. Proper cleaning of eating and cooking equipment, correct disposal of garbage and human waste, and protection of utensils from flies can all help prevent the spread of illnesses. All patrol members must regularly check each other for signs of any injury, however slight, as dust and insects can cause infection of minor cuts and scratches.

Feet, which often provide the SAS soldier with his only source of mobility, require particular attention when operating in the desert. They tend to overheat when they are encompassed in standard combat boots. Therefore desert boots, complete with suede uppers,

Right: Weighed down with a heavy backpack, an SAS trooper descends a rock face. His skill at rappelling ensures an effortless descent. Note the wearing of gloves to prevent rope burns.

rubber soles and ankle support, are often adopted by members of the Regiment in preference to the issue item.

Vehicles used by the SAS in the desert, such as long-range Land Rovers or Light Strike Vehicles (LSVs), are specially designed and modified to cope with the harsh climate. The conditions encountered demand more maintenance than that required in temperate climates, vapour locks occur in fuel systems, batteries are more prone to failure, and fuel, oil and coolant filters get blocked by sand and dust. Tyre inner tubes have a tendency to rupture, despite the fact that the pressure is kept on the low side to cope with travelling over soft sand. Extra inner tubes, additional spare wheels, air pumps and repair kits, as well as large quantities of water

Below: A soldier wearing NBC kit in the desert can quickly succumb to sunstroke and exhaustion.

Above: SAS desert uniform, circa 1991. Headgear is essential for protection against the sun.

and ammunition, all increase the load each vehicle must carry to remain mobile.

The effects on weapons and ammunition caused by sun and sand must also be considered by SAS squadrons operating in desert terrain. The need to protect them from heat and dust means that weapons and ammunition must remain covered when not in use. A clogged barrel, for example, can lead to an inborne detonation. After a rifle is cleaned, for example, all oil is removed from the moving parts, as contamination of the lubricant with sand results in an abrasive paste which damages them. SAS foot and vehicle patrols can protect their various personal and vehicle-mounted weapons by using commercially-available stretchable plastic sheeting (such as cling film), or condoms, to keep out the dust.

As high temperatures also effect ammunition, it should remain under ventilated shelter even when boxed. As a rule it is safe to fire if it can be held by bare hands. The desert heat can also have an adverse affect on radio equipment. Solder is likely to become brittle and connections may become loose. In addition, batteries require re-charging more frequently — all extra back-breaking work for the patrol signaller.

Camouflage nets are important for concealing individuals and vehicles

Camouflage is even more important in the flat, featureless terrain of the desert than it is in temperate climates, where natural cover is more readily available. Camouflage nets are important for concealing both individuals and vehicles, and should have hessian or cloth sown into them to provide shade from the sun. They also provide some protection when using optical equipment, such as binoculars or telescopic sights which, if the lenses catch the light, can reflect it up to a distance of 10km. Anything glass, or with a gloss surface, needs to remain covered up during the hours of daylight for similar reasons.

Positions such as harbour areas need to be camouflaged as soon as they are occupied, and observation posts (OP) require close attention to conceal them from the enemy. Natural cover such as caves and dead ground provide protection from both the elements and prying eyes, but deep desert wadis should be avoided, except as a last resort. Although normally dry, the desert is prone to sudden heavy downpours. Such rains cause flash floods in wadis and troops, even special forces (SF) soldiers, have drowned in the desert.

Below: Placing plastic bags over tree branches (a), plants (b) and cut leaves (c) produces water.

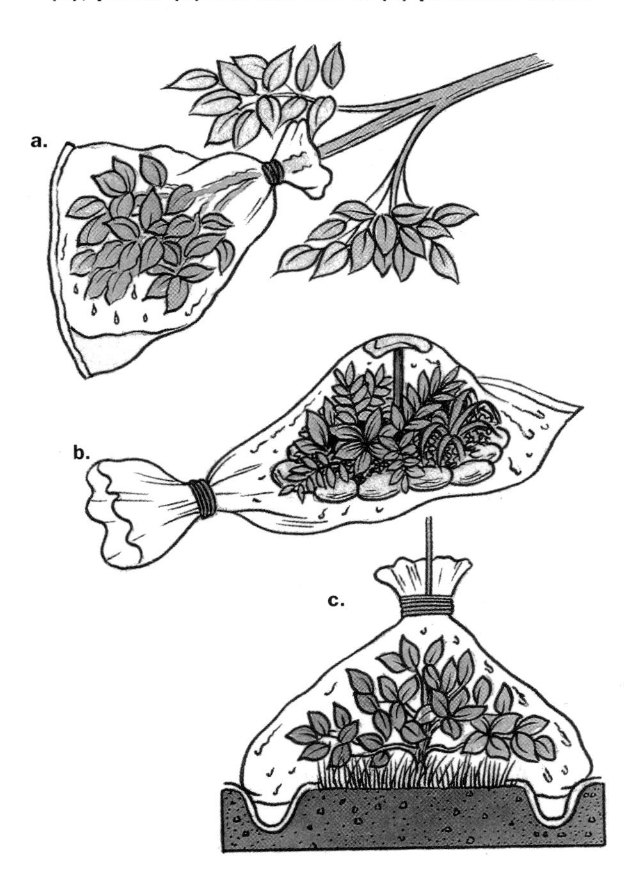

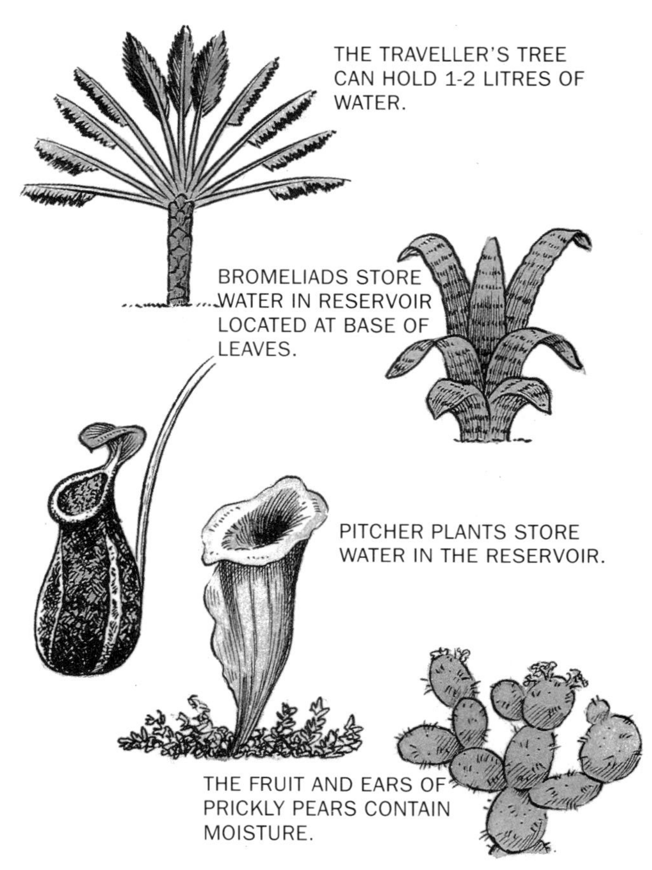

Above: Various cup-shaped plants and the cavities between the leaves of bromeliads hold water.

However, it is the absence, rather than the abundance, of water that is generally a problem. Both water and food in desert regions are scarce, and so SAS patrols have to carry their supplies with them. The use of air drops to re-supply long-range patrols is not without risk, but unless the patrols have access to pre-cached stores it is often the only reliable method of re-supplying them. One advantage of the desert is that the relatively flat terrain affords the opportunity of constructing temporary air strips, which can be used by fixed-wing aircraft to deliver supplies, as well as extracting wounded.

The buggy-like LSV makes short work of flat, compacted areas of sand

In emergency situations troopers may have to rely on their combat survival training and construct a desert still to obtain enough water necessary to sustain life. These stills are relatively simple to build and can provide up to a litre of drinkable water over a 24-hour period. The chosen site should be a place where water would normally collect. A hole one metre deep and two metres wide should be dug and filled with old vegetation, waste water, urine, etc. A container should then be placed in the centre, together with a drinking tube extending to the side, and the hollow should be covered with plastic sheeting with a rough under-surface.

The sheeting is then secured by stones around the circumference, with one or two placed in the centre to depress the plastic directly over the container. The best time to construct a desert still is during the day, and the optimum time to collect the condensation that forms beneath the plastic and drips into the container, is early in the morning. As certain desert plants can also provide water, SAS soldiers are instructed in the local vegetation as part of their initial briefing on arrival in a particular theatre.

Movement across desert by vehicle can be either slow or rapid, depending on the type of local terrain. The buggy-like LSV, with its sturdy construction and low centre of gravity makes short work of flat, compacted areas of sand and dunes, but rocky areas must be approached with more caution and slower speeds.

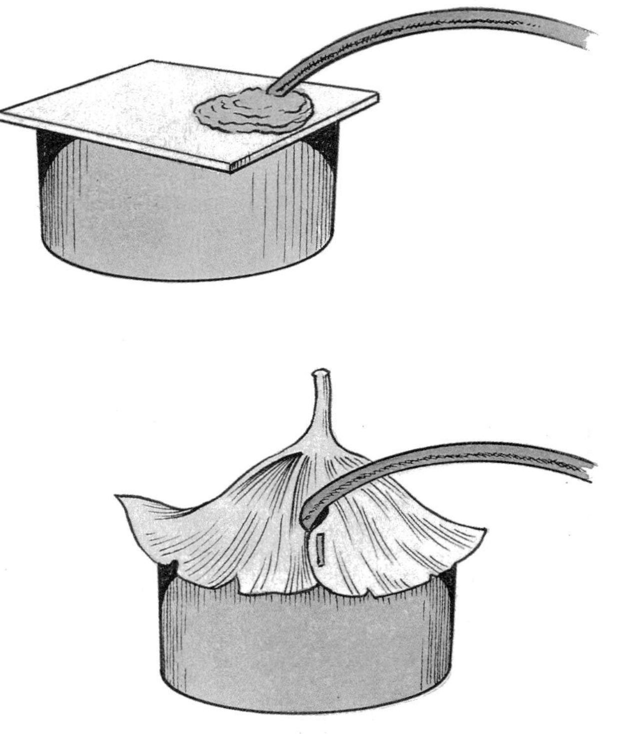

Above: Distillation. Vapour from liquids boiled in covered vessels can be drunk after they cool.

Lack of cover, coupled with the creation of dust clouds during the dry season, means that most vehicle movement is conducted at night. Accurate night navigation skills are especially important in the desert, as tracks marked on maps are often inaccurate, and behind enemy lines they will be used by the opposition.

SAS patrols are conducted by Land Rover vehicles in the desert

Dead reckoning is the most common method employed, with the lead vehicle driving along a series of pre-planned bearings, while continually checking the distance travelled. Enemy positions and areas of hostile activity are usually approached/infiltrated by British Army patrols on foot during the hours of darkness, but SAS patrols, with the exception of close-target reconnaissance (CTR) teams, are conducted by long-range Land Rover vehicles in the desert, just one more example of how the specialist skills of the SAS are used to conquer the hostile desert environment.

TROPICAL REGIONS

The ultimate success of any SAS campaign or operation depends largely on the skills of individual troopers, noncommissioned officers (NCOs) and officers. It is for this reason that Britain's SAS Regiment sets great store by its ability to select the most suitable men from the many candidates who volunteer for service with Britain's primary SF unit. These men undergo very stringent selection procedures, most of which include land navigation exercises over difficult terrain, and under conditions of severe physical and mental stress. It is interesting to note that the areas chosen for these exercises include some of the most inhospitable mountainous terrain in the British Isles, and ultimately the SAS Regiment's traditional proving ground: the jungle, a region that has almost become a 'second home'.

Jungles are characterised by high temperatures, frequent, very heavy rainfall, and high humidity. Primary jungle contains trees that can grow to a height of over 60m and forms a thick canopy of leaves so thick that very little light reaches the jungle floor. There is little undergrowth to prevent movement, though the poor light means visibility is reduced to 50m, or less. Secondary jungle occurs when sunlight reaches the jungle floor, as on river banks and where primary jungle has been cleared by man. Here the ground is covered by grasses, ferns, vines and shrubs, which can reach a height of three metres. Visibility is poor and movement is slow and tiring, often requiring much heavy work with a machete.

Below: A solar still constructed in the desert. Droplets of water run down the roughened underside of the plastic sheet and collect in the can.

Individuals will sweat a lot, especially if carrying heavy loads, and water, although available in plentiful quantities in most areas, is often contaminated. In addition, disease and parasites thrive in a jungle environment, and so soldiers operating in it need to guard against infection and animal threats. Like the desert the jungle includes a wide variety of terrain but, whereas the former is largely covered by dust, sand and rocks, the latter is mostly hidden by an impenetrable jungle canopy. The sub-tropical rainforests of southeast Asia, for example, contain a vast multitude of mountains, valleys, rivers, streams and swamps. Out of all the environments in which an SAS trooper might be tasked to fight, the jungle is perhaps the most simple in which to survive, as long as he has been properly trained.

Providing the trooper has become properly acclimatised, has an adequate knowledge of where and how to obtain food and water, and how to construct rudimentary shelters using local materials, he should have little problem remaining fit to fight. Physical fitness and the ability to remain healthy are perhaps the most important attributes in the jungle. Wading through swamp for hours at a time, for example, carrying a weapon over his head, is a severe test of a trooper's physical endurance. In order to overcome the prevalent sickness and many diseases associated with the heat and humidity of a jungle environment, soldiers need to be aware of the effects of the climate and maintain a high standard of personal hygiene.

Mental preparedness is another important factor, and troopers need to overcome a somewhat natural fear of their dank, dull and claustrophobic surroundings. The climate and conditions can cause lethargy and mental apathy if the men are unprepared, and there are some soldiers, although few of them belong to the SAS, who will never be able to adapt themselves to operating in what is a totally alien environment.

One of the most important aspects of jungle craft is the ability to locate and prepare water. As in the desert, a soldier's daily water requirement is far greater than that required to sustain life in a temperate climate. During their initial period of acclimatisation soldiers may need up to seven litres per day, plus additional salt in tablet form, just to remain fit and healthy. Despite the fact that they may be in an area where they are constantly soaked to the skin, obtaining good quality water presents problems. In the tropics it is vital that all water used for drinking and cooking is filtered and sterilised before being consumed, boiling alone is ineffective in removing bacteria.

Chopping through jungle is tiring, noisy and leaves signs of passage

Although water is usually available from rivers and streams, there are some areas of the jungle where it has to be obtained by other means. Collecting rain water from natural hollows or tapping certain plants are two of the most common methods used, though a jungle still, similar to that used in the desert, can be built in areas where sunlight penetrates the jungle canopy. A device known as a Millbank bag, a special canvas water container employed by the SAS since Malaya, is used to filter all collected water, which is then sterilised.

During their acclimatisation period all troops entering a jungle environment receive instruction in the skills required to survive and fight in the *ulu*, as the jungle is known to members of the SAS. However, other jungle skills and drills need to be mastered before the men are ready to take on and beat the enemy. Engagements with enemy forces in the jungle are usually conducted at very close ranges by small groups on each side. The ability to defeat the enemy in such a situation depends on the ability to move silently, to detect the enemy first, and to shoot accurately and quickly at moving targets.

As the enemy will also be experienced in operating in the jungle, camouflage, both of the individual and his personal equipment, is especially important. The ability to see through the enemy's efforts to camouflage himself, and to achieve total surprise in ambushes and target attacks is also crucial. Sound, and indeed the absence of sound, can easily give away a patrol's position, so total adherence to noise discipline procedures are vital, especially when moving through dense jungle when the enemy may be close. Lack of the normal sounds of the jungle, such as birds and insect life, can alert an enemy just as much as the rattle of mess tins or the creak of wet, heavily-loaded webbing equipment. Camouflage, navigation and tracking skills, learned during initial training in Britain, are therefore greatly developed after the troopers arrive in-theatre.

Careless movement, leaving tracks or signs, and creating sounds or smells are all things that have to be avoided if the SAS soldier is to escape detection. For instance, cooking should only be carried out when there are no enemy in the vicinity, and then only in a confined container. In addition, the traditional Hollywood image of troops hacking their way through thick jungle is far removed from the SAS reality. Chopping through jungle is tiring, noisy and leaves signs of passage. It makes movement slow and reduces reaction time as personal weapons cannot be held at the ready. The point was stressed by Lofty Large, a veteran of the Borneo campaign: 'Smashing through the sticks like a rampaging bull elephant gives ample warning of your approach, ensuring a hot reception.' Cutting down foliage should only be done in a survival situation, or when creating a clearing or constructing a shelter.

One of the most important specialist items is footwear

Apart from the requirement for special skills, specialist clothing and equipment are also needed for jungle warfare. Without the proper equipment, coupled with the knowledge of how to use it, the jungle is a highly dangerous environment in which to operate. One of the most important specialist items is footwear. The feet

Above: In Kuwait a US helicopter throws up a cloud of dust as it prepares to land in the desert.

and legs must be protected at all times, not least from snakes, scorpions, leeches, chigoe and centipedes. The first issue jungle boot was made from rubber and canvas, and was worn by the original SAS jungle fighters during the Malayan campaign. It was not ideal, however, and so, during the campaign in Borneo, the US issue jungle boot was adopted by members of the SAS. This lightweight boot was designed with the jungle in mind, and incorporated lessons learned from the wars in southeast Asia. It has a heavy moulded sole capable of coping with thick mud and providing excellent grip, a protective leather base with ventilation holes to allow the foot to breath and water to escape, and strong canvas uppers to protect the ankle and lower leg. In Borneo the soles of troopers' boots were often re-cut to leave the pattern of an Indonesian, as opposed to a British, foot-print on the ground, thus deceiving the enemy.

In Malaya and Borneo the olive green (OG) shirt and trousers gained favour with the Regiment and, like the jungle boots, were often worn in other operational theatres. The shirt had long sleeves and full-length tails manufactured from 100 per cent cellular weave cotton material, with heavier duty cotton trousers. The latter were high waisted with a cross-over buckle front and button fly. The shirt tails and waist were effective in protecting the wearer from the ever-present disease-carrying insects and leeches, as well as palms, atap, rattan and bamboo which can inflict nasty deep wounds on bare flesh.

Over the last decade OGs have been gradually replaced by DPM camouflaged jungle clothing which, like its predecessors, is made from quick-drying cotton. Constant humidity in the jungle can cause clothing, and indeed skin, to rot within a relatively short space of time, so it is important to dry clothing and

Above: Dust and sand can damage working equipment, as these SAS troopers have discovered.

body out whenever possible. Body temperatures can drop dangerously low when a man is exposed to continuous rain, or is operating at high altitude; getting dry and staying warm can mean the difference between life and death. One of the most effective ways to do this is to get under shelter.

There is ample material for building emergency or temporary shelters in most tropical regions. Bamboo, palm fronds and rattan can be used to construct quite elaborate shelters, while vines, creepers and dried fern stems can be used to hold them together. Issue shelter sheets, ponchos and parachute cord are also used by troopers operating in the jungle. In Borneo, for example, SAS hides often consisted of a waterproof sheet covered with jungle leaves. Careful siting of shelters in tropical theatres is especially important, and where possible they should be constructed in thick cover which offers more than two escape routes out. They should be erected in an area which provides water, though not one where flooding is likely, and must be away from any overhead cover which might come crashing down during the night.

Because of the abundance of food available in the jungle, sustenance is less of a problem than in either temperate or arctic environments. Although resupply by RAF aircraft, either dropped by parachute from a C-130 Hercules aircraft or ferried in by helicopters, can supplement the SAS seven-day patrol ration pack, the jungle itself contains a wide variety of fruits, roots and leaves that are edible. In addition, there are also deer, pigs, snakes, monkeys and birds in many areas of jungle. SAS troopers in Borneo supplemented their rations with curried snake and monitor lizards, fresh fish and monkey. Less appetising, but prolific and nutritious, are the snails, slugs, grubs and other

Above: Split bamboo stems can be use to construct jungle shelters. Note the drainage can.

insects which form a substantial part of the diet of the indigenous inhabitants, and have on more than one occasion sustained members of an SAS patrol short of supplies.

Movement in the jungle can be both agonisingly slow and dangerous

Logistics, at least as far as food is concerned, is less of a problem in the jungle than in a temperate climate, but the dangers to weapons and equipment — from humidity, mud, and rain — are greater. They therefore have to be stripped and thoroughly cleaned every night to keep them in working order. Movement in the jungle can be both agonisingly slow and dangerous. Limiting factors include the density of foliage and the associated difficulties in navigation, especially in the poor light of primary jungle. Enemy ambushes mounted along likely trails and on crossing-points are an ever-present hazard that can only be overcome by selecting an obscure route, often over difficult terrain. A high level of movement discipline, coupled with the ability to navigate accurately along compass bearings, are all important aspects of jungle operations.

Drills, such as immediate actions (IAs) on encountering ambushes, have to be especially quick and well-rehearsed. Most patrol movement is in single file, making it open to attack from any direction. Anti-ambush drills must therefore be simple, immediate and aggressive. Obstacle-crossing procedures, especially the selection of 'safe' crossing points, help reduce the risk of ambush. Those that are obvious are normally ignored by SAS patrols, unless secured by friendly forces.

Grenades are not carried on the chest as they present a target for the enemy

General patrol rules when moving through the jungle include each man being equipped with one magazine full of tracer bullets, which can be used to identify enemy positions in a firefight. Magazines in ammunition pouches are usually placed upside down, with the bullets pointed away from the body. This keeps dirt and water out of them and, if the rounds go off after being hit by enemy fire, they will not injure the wearer. Grenades are not carried on the chest as they present a target for the enemy, who will shoot at them to inflict several casualties with one strike.

One of the most effective methods of moving through the jungle, as opposed to being transported above the canopy by helicopter, is by river. In many areas of the tropics rivers are used in the same way as roads. However, using the waterways is not without danger, not only from enemy ambush and being spotted from the air, but also from disease. Leptospirosis is one disease prevalent in the tropics, and one which SAS soldiers operating there have to guard against. Spread via the urine of infected animals, leptospirosis contaminates the water in streams and rivers. As a precaution all water is thoroughly sterilised before consumption.

The threat of contracting mosquito-borne diseases, such as malaria and dengue fever, can be minimised by immunisation and taking measures to prevent insect bites. Heat-associated illnesses, such as heatstroke, exhaustion and heat cramps can nearly always be prevented by acclimatisation, fitness and the correct fluid intake, while skin diseases can be cut down by proper attention to personal cleanliness and general hygiene.

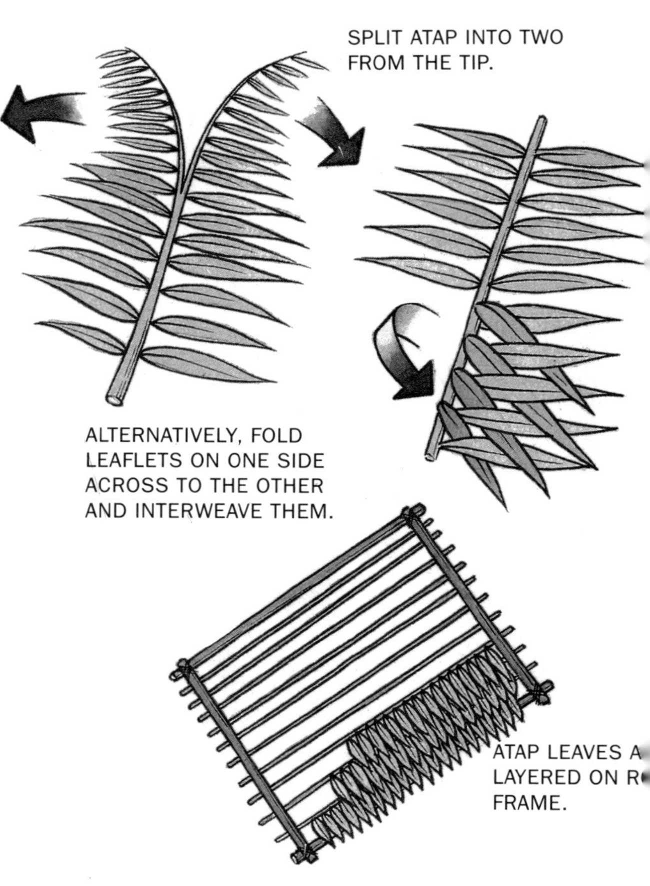

Right: Found in tropical regions, atap is extremely useful for constructing temporary shelters. The leaves can be either split into two from the tip or interwoven as shown. The vine must be handled with care as the barbs on the tip are very sharp.

The jungle remains an environment where the careless die and the careful survive; the fact that few SAS soldiers have fallen seriously ill or died from preventable diseases during the Regiment's long association with the *ulu* is testimony to their high standard of training and professionalism.

POLAR REGIONS

Extremes of cold, together with high wind speeds, are the characteristics of polar regions. The SAS Regiment's soldiers receive intensive training in arctic warfare, an exacting area of combat and survival where almost any mistake can kill quicker than it would in the desert or the jungle. In polar regions the body can only operate effectively as long as its temperature is kept within certain limits. Intense cold strikes the mind as well as the body, increasing reaction time and requiring longer periods for even the most simple tasks. The wind can freeze exposed flesh in minutes, and immersion in cold water can mean death for even the well-protected in under 20 minutes.

Cold weather effects all aspects of the soldier's daily routine: the clothing he has to wear, the specialist equipment he must take with him, the food he has to eat to maintain his strength, and the type of shelter he will have to construct to protect himself from the elements.

Cold weather presents a physical challenge to the soldier, as well as his equipment. Fitness and personal hygiene are important, though a balance has to struck over the latter. Shaving daily, for example, is often discouraged as it removes oils that protect the face from frost-

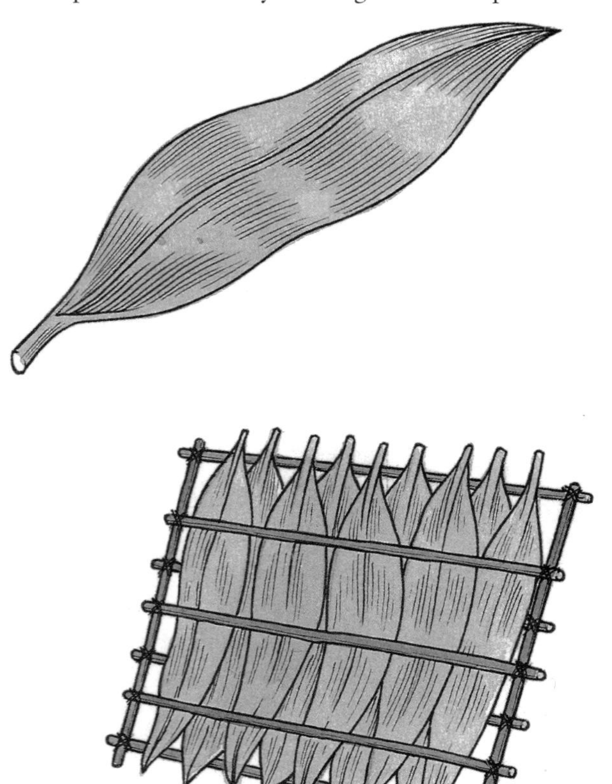

Left: Elephant grass (shown here) and other large tropical leaves are useful for quickly constructing shelters as only a small number are required.

Above: Being able to accurately navigate in a jungle environment is essential on operations.

bite. Most cold injuries can be prevented by applying sensible precautions, but trench foot and exposure can be a problem if soldiers have not got access to proper food and shelter. Severe cold can cause frostnip, superficial or deep frostbite, and shock, and for the two latter conditions immediate casualty evacuation is the only option to avoid death.

Clothing has to be tough, light, windproof and waterproof

The need to cover the whole body, especially head, hands and feet, is of paramount importance when operating in extreme cold weather conditions. Clothing has to be tough, light, windproof and waterproof; it also has to provide insulation down to -40 degrees F, while remaining suitable for work in temperatures above zero. Natural fabrics, such as wool and fur, are suitable for cold temperatures as they trap air and maintain a barrier of warm air close to the body. However, the most effective and flexible system for the required temperature range operates on the simple 'layer' principle, which allows for layers to be added or removed depending on the temperature and level of activity. Several layers of light- to medium-weight clothing are far more effective than one heavy garment, even when this single item might consist of a number of layers itself. By using the layer method clothing can be adjusted to suit the situation, and body heat maintained at the correct level.

The SAS Regiment has tried and tested much of the cold-weather clothing now in service throughout the British Army, including the windproof smock and trousers. These lightweight, closely-woven cotton garments are designed to protect the wearer from the worst of the elements, especially wind and rain. The current DPM versions replace the more traditional 'Dennison-type' camouflaged smock and trousers, initially adopted by members of the SAS in World War II. Loose-fitting, these quick-drying cotton gaberdine garments have large pockets fitted with big buttons, allowing their use while wearing gloves or mittens.

Windproofs can be worn alone or in conjunction with outer garments, such as two-piece waterproof suits. The SAS pioneered the use of protective clothing manufactured from Gortex fabric, an expensive material designed to keep rain out while allowing moisture to escape outwards. Traditional waterproofs were well-known for trapping moisture, such as sweat, inside them, with the result that the wearer would become damp quite quickly if involved in physical exercise. Gortex garments were initially adopted for those soldiers manning covert OPs, where men might have to remain exposed and inactive for long periods. They are especially useful in an arctic environment, where they prevent rain from entering and provide protection against melting snow.

The SAS arctic warfare rig includes bergen , snow shoes and skis

Thermal liners, comprising a green nylon quilted jacket and trouser set, are sometimes worn beneath windproof clothing to supply additional warmth, especially when the soldier is not involved in hard physical activity. They can be worn with, or without, outer garments depending on the level of protection required, and over, or instead of, long woollen/cotton mix issue underwear. Silk 'long-johns' and vests are sometimes purchased privately by SAS soldiers operating in cold weather conditions. These are generally reckoned to be the lightest, most comfortable and effective layer that can be worn next to the skin.

A range of mittens and gloves are also available to troops working in the arctic, and include inners and outers of both types. The standard set for manual work and moving in sub-zero temperatures consists of an inner mitten, which is worn beneath a windproof and waterproof outer one. These are often replaced by water-resistant gloves when working in wet conditions for short periods, such as when constructing snow shelters or igloos. Wristlets or wristovers can be worn beneath gloves or mitts to further reduce heat loss and help keep hands and fingers warm.

Proper headgear is vital in the arctic, where it is estimated that up to 70 per cent of an individual's body heat can be lost through an uncovered head. Various articles are on issue, including the arctic cap with fold-down ear flaps and the standard issue GS cap, both of which offer protection against the wind and the cold. Hoods can also be worn over headgear, although SAS soldiers have a natural aversion to wearing anything that might interfere with their hearing.

Footwear depends on the particular mission and the type of terrain which is to be traversed. The issue ski-march boot is fairly basic, and so most SAS mountain and arctic warfare specialists prefer the more comfortable German *Bundeswehr* ski-mountaineering boot. The latter is tougher and heavier, offers better protection from the elements, and gives greater support when skiing.

The SAS arctic warfare rig also includes bergen rucksack, snow shoes and cross-country skis. The standard SAS bergen is sometimes replaced by the 'arctic bergen' which has a light tubular aluminium H frame, allowing the sac to be carried high on the back, the most effective position for ski-mountaineering. The 'arctic bergen' has straps and attachment points for skis and poles, and with an adjustable padded waist belt is more comfortable than the standard-issue bergen.

Snow shoes, generally a small type, are used for close quarter work

Britain's 3 Commando Brigade, Royal Marines, is the unit generally regarded as the country's foremost mountain and arctic warfare troops. However, much of their personal equipment, including their skis, is outdated. The SAS, like the Marine Commandos' special forces, the Marine and Arctic Warfare Cadre, does not have the same restrictive budget as more regular formations. It can therefore afford better equipment, which is often provided free by manufacturers for testing. As a consequence, the skis used by SAS soldiers are often much

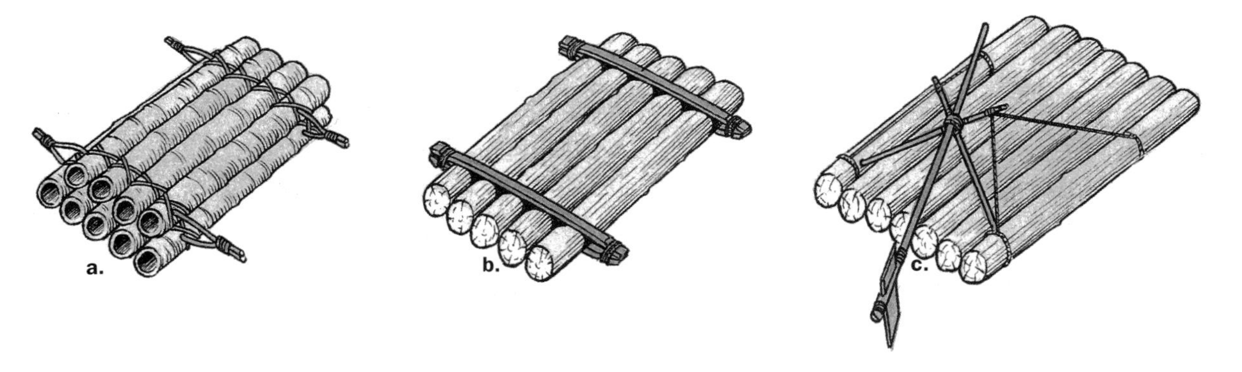

better than the standard British Army wooden 'plank' with cable bindings. The most common type of SF ski is the Alpine tourer/downhill with ski-mountaineering bindings. These allow the heel to be clipped into the binding for downhill stretches, released for cross-country journeys, and raised for uphill climbs.

Snow shoes, generally a small type, are used for close quarter work, such as a target attack, where skis would be unmanageable, or daily routine tasks around a field location which is covered by deep snow. Not all movement in the arctic is conducted on foot or skis; the SAS make extensive use of both helicopters and over-snow vehicles when the tactical situation

Top: In the jungle rafts can be made from the abundance of local materials available. Shown here are a bamboo raft (a), a gripper bar raft made from logs and four stakes (b), and a raft equipped with a rudder mounted on an A-frame (c).
Above: An SAS team in Malaya, 1954.

Left: Drenched in sweat, an SAS soldier moves cautiously through the jungle in Borneo, constantly guarding against the threat of an enemy ambush.

allows. Each year the Royal Navy deploys two helicopter squadrons to the arctic in support of Britain's contingent to NATO. The squadrons' highly-skilled aircrew, whose expertise was proven during the Falklands campaign, provide the SAS Regiment's Northern Flank forces with helicopter support.

Low temperatures can make all movement difficult. Bad weather and poor visibility can often keep aircraft grounded for days at a time, and can also restrict surface traffic. In addition, maintenance of vehicles and weapons is extremely difficult in sub-zero temperatures. In extreme cold weather conditions normal temperature lubricants have a tendency to thicken, in weapons this can cause stoppages and sluggish actions. Weapons therefore have to be stripped completely and thoroughly cleaned to remove all lubricants. Only the camming surfaces of the bolt are lightly lubricated, the rest of the weapon is left dry. Snow and ice must be kept out of the working parts, sights and barrel. A weapon that is taken from the outside into a heated shelter, for example, will 'sweat' for up to one hour. After this period it must be thoroughly cleaned before it is taken outside again. If this is not done the condensation will freeze once it is outdoors. During firing the hot parts of the weapon must not come into contact with snow or ice, as the snow will melt and, once the weapon is cool, will form ice.

In the extreme cold certain radio components can become brittle

Ammunition must likewise be cleaned of oil, and all ice, snow and condensation must be removed. Stored ammunition must be left in its boxes, raised off the surface and covered to protect it from the elements. In the extreme cold certain radio components can become brittle and fail, battery strength is reduced and condensation is an ever-present problem for all general electrical equipment.

The need to get out of the wind and create a warm atmosphere is the key to arctic survival.

Above: An SAS soldier with his arctic warfare kit.
Right: In polar regions all movement is slow.
Below: Snow caves can be used as shelters in open areas containing deep and compacted snow.

RATIONS, DIGGING TOOLS, ETC
RUCKSACKS AS PILLOWS
SNOW BLOCK
STOVE
BOUGHS OR OTHER INSULATION

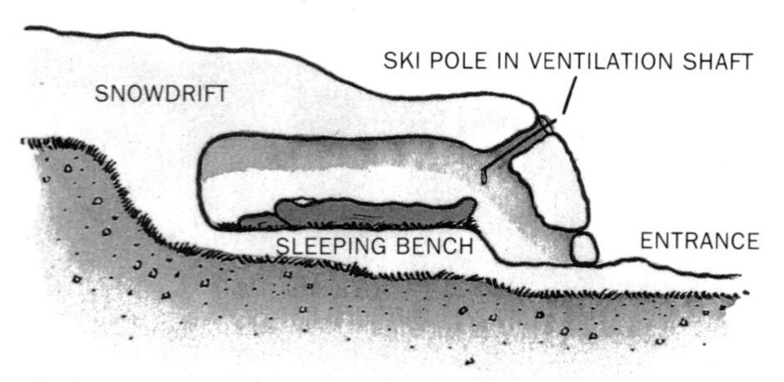

Skills used by Eskimos and Lapps for many hundreds of years have been adopted and developed by the modern arctic warrior. Although tents and brushwood bivouacs can be erected above ground and do offer protection from the elements, perhaps the most effective constructions for the soldier operating in an arctic environment are snow shelters. As long as the snow is deep enough, various types can be built with relative ease, not only providing a warm place to rest, but also a well-concealed tactical position. Snow shelters are far easier to camouflage than tents or brushwood 'bivvis' and are much more comfortable.

The rules governing the construction of snow shelters are simple: the entrance must be lower than the sleeping platform; the arched ceiling should be smooth to prevent condensation dripping on the occupants; a shovel should be kept inside in case of the roof collapsing; and there should be a permanently open ventilation hole. Aside from these rules, the style of an individual shelter depends on the time and imagination of its builder.

Snow holes are simpler structures, being basically burrowed into the side of snow drifts.

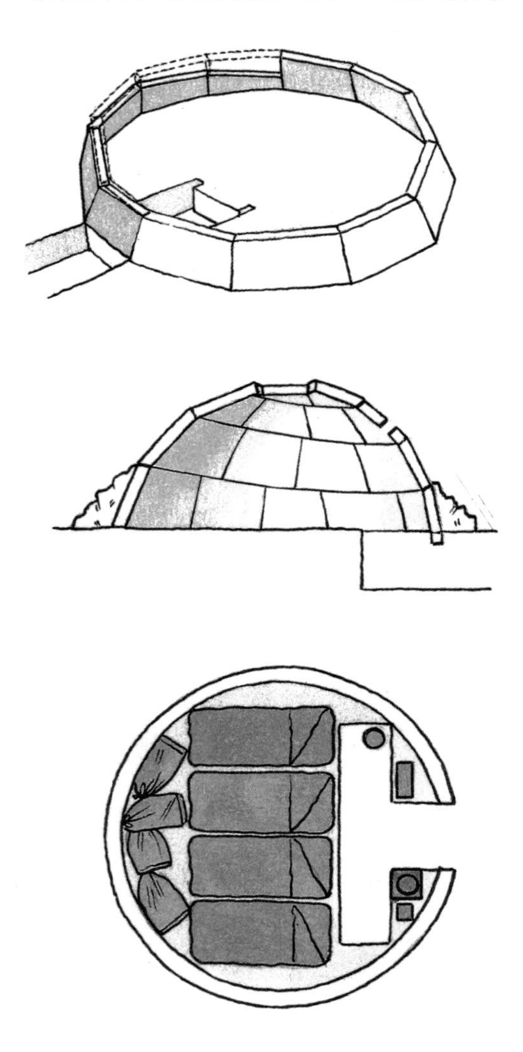

Above: The igloo is the ideal arctic shelter, being soundproof, wind-resistant and large enough to sleep a four-man patrol in reasonable comfort.

Snow caves can be constructed using drifts but, instead of exiting through a tunnel, snow blocks are carved after the inside has been dug out, and these are then used to seal the cave. Snow trenches can be dug if no drifts are available, but these tend to be less comfortable and rather cramped. By far the best design is the igloo.

The igloo is stronger, warmer and provides better protection from the elements than any other type of snow shelter, but to build one requires a certain amount of practical experience combined with good quality, compacted snow. Igloos can vary in size and be made to house up to ten men. They are constructed from snow blocks built up in a circle around a floor area dug-out to ground level. The blocks are built up, and towards, the centre, with each block being shaped from the second layer upward so that the wall spirals inwards as it rises. The final stage involves placing the key

stone, an irregular shaped snow block, into the opening at the top. This will prevent the structure from caving in, and is passed down from the outside and positioned from the inside. The interior should be finished off by plugging any cracks and holes with snow. A tunnel should be built below the surface, if the snow is deep enough, but above if not, and the entire exterior should be left rough. This will allow fresh snow to adhere to the finished structure more easily, thus aiding the overall camouflage affect.

Below: Two troopers 'tabbing' across an arctic environment and dragging a heavily-laden *pulk*.

There is no shortage of water in the arctic, although snow and ice have to be melted to get it, which requires large quantities of fuel. A surprising amount of drinking water is needed by troops operating in the arctic — up to five litres per day — depending on the level of physical activity. Cutting through ice to get water from running streams and lakes is the most economical method of obtaining the necessary amount. Food is available, although hunting and fishing can take up a soldier's valuable time. It is far better to obtain rations and fuel from pre-positioned caches, with additional supplies brought in by over-snow vehicle, boat, helicopter, para-drop, or hauled on a sledge.

During the winter months almost all cross-country movement in the arctic is restricted to specialist vehicles or skis. All SAS soldiers who deploy to the arctic for exercises receive in-theatre ski training on arrival as a part of their arctic warfare (AWT) and arctic survival training (AST). In a two- to four-week training programme troopers learn snow-shoeing, fast military skiing, trail-breaking and ski-joring.

Below: A Royal Marine recruit on exercise in Norway. The need to keep warm and dry is imperative when operating in such climates.

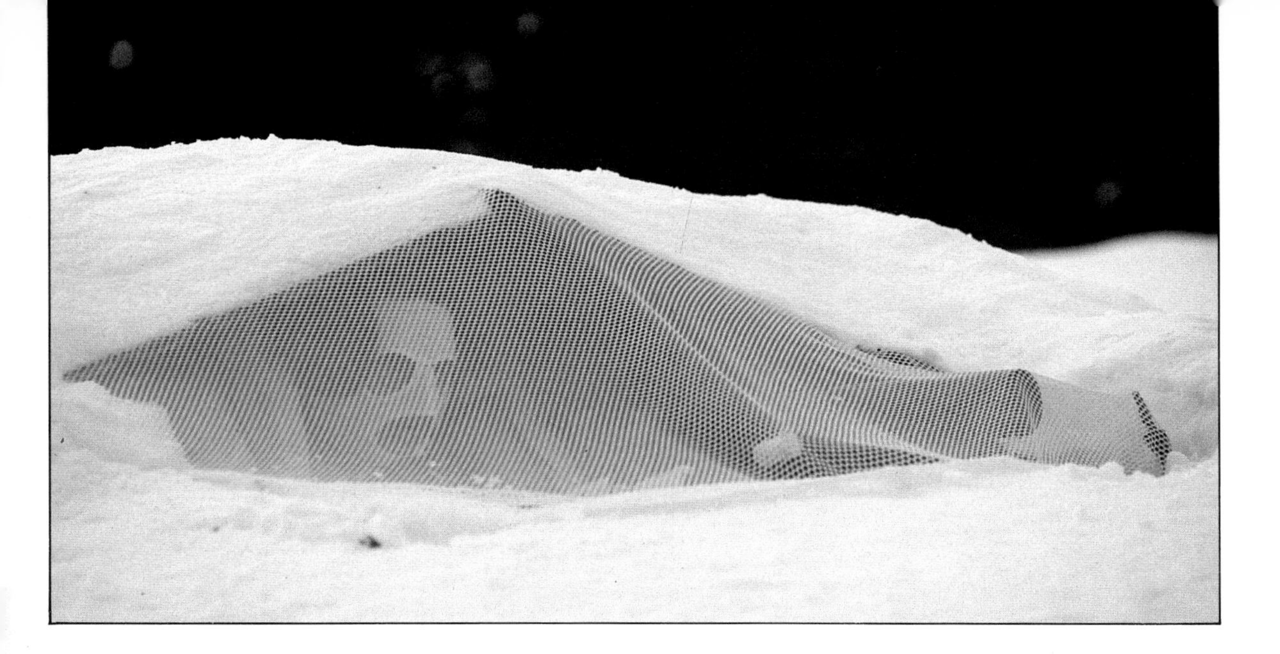

Above: An SAS hide in the snow. All troopers are well-practised in the art of arctic camouflage.

Military skiing is a faster and more practical way to cover ground in the arctic than using snow shoes. More comprehensive than civilian sport, downhill and *langelauf* skiing, it teaches soldiers to ski across all types of terrain in all weathers. Although snow and weather conditions can greratly influence speeds, a trained SAS ski patrol, equipped with weapons and bergens, would normally expect to cover seven kilometres an hour over reasonably flat terrain. Faster speeds can be achieved by ski-joring, a technique by which ski troops are towed behind cross-country vehicles, such as the tracked over-snow vehicle. By this method speeds of up to 30km/hr can be achieved, depending on conditions and terrain.

A patrol will cut a track through the snow following a breaker

Trail-breaking is a technique by which reconnaissance troops, such as SAS teams or units operating in the advance of the main force, mark a track or trail for those following. A patrol will cut a track through the snow following a breaker, who moves through virgin snow as directed by the patrol commander. He is followed by two cutters to his left and right, who remove obstructions from either side of the trail. Behind them come the packers, who will flatten the ground and fill in the depressions to make the going easier for the troops who will follow.

Accurate navigation for all SF troops operating in arctic conditions is vital. It is easier to become disorientated in the arctic than in other theatres as the terrain is often inadequately mapped and can actually change with the weather. Getting lost can be fatal, and so the need for good compass work and navigation skills becomes extremely important for the individual SAS soldier. Prismatic compasses, for example, need to be kept warm to stop them becoming sluggish or freezing-up entirely. Magnetic variation can fluctuate wildly and the stars can be invisible for days at a time if the weather deteriorates.

The arctic is the most hostile environment in which the SAS soldier is likely to train. In mountainous regions within the arctic circle the trooper faces all the dangers of the mountains, plus the adverse weather. Deserts and jungles can be easy by comparison. In the latter a soldier can survive for longer periods without shelter and, should he become separated from his colleagues and equipment, he could survive for several days. In the arctic, by comparison, the expected life span in such circumstances is hours or less. A soldier cannot afford to make any mistakes. However, as with other theatres of operations, the arctic is yet one more environment which has been mastered by the troopers of the Special Air Service.

204

SPECIALIST INSERTION METHODS

Dropping by parachute from a height of 10,000m, racing across the desert in heavily armed fast-attack dune buggies, rushing towards the shore in nimble Rigid Raider assault boats or following the terrain in a specially converted long-range Chinook helicopter — whatever the mission the Special Air Service can call upon a wide variety of special-to-purpose equipment, vehicles and other craft to enable its men to be covertly inserted into an area of operations.

Britain's SAS soldiers were among the first Allied troops to be deployed in the Gulf theatre of operations during the conflict against Iraq. Once again, the SAS Regiment's experience in conducting behind-the-lines missions proved vital to the overall success of the Allied operation in liberating Kuwait. Reconnaissance, intelligence gathering, pin-pointing enemy targets and attacking vital installations were all highly-specialised and dangerous tasks, and among those regularly undertaken by the SAS and their US and Allied counterparts.

There can be little doubt that behind-the-lines operations carry the greatest risk of death, injury or capture for SAS personnel. Working within enemy-occupied territory,

A C-130 Hercules heavy transport aircraft carries out in-flight refuelling. Air-to-air refuelling allows special forces teams to be inserted deep inside enemy territory.

particularly if security is strictly enforced, or actually inside his home country where it is often extremely difficult to blend in with the local 'scenery', makes any mission by special forces (SF) extremely hazardous. In operations of an aggressive nature, such as close-range target attacks, raids and ambushes, the need for speed and surprise are of paramount importance. Alternatively, in missions of a more 'passive' nature, such as reconnaissances, it is vital that patrols remain undetected, and secrecy is considered more important than speed. In both types of operation, however, success or failure can often hinge on the means by which troops are inserted into and extracted from their area of operation.

The assault group was landed almost on top of the target

The method by which an SAS patrol is taken to, and removed from, enemy occupied territory is determined by its specific mission. In a raid demanding speed rather than stealth the attack force might be inserted extremely close to the target. There was at least one such case reported during the war against Iraq. A combined Allied (US/British) strike force was tasked with capturing an Iraqi surface-to-air (SAM) missile and launch system and returning it to friendly lines for analysis. The assault group was landed almost on top of the target, and relied on speed and surprise to secure the objective. The method of insertion and extraction chosen in this case was by helicopter, using machines borrowed from the Egyptian military. These were painted in Iraqi camouflage in a successful attempt to confuse the enemy, a ruse that proved so convincing that their return gave rise to the rumour that several Iraqi helicopters had defected. Allied personnel briefing the media refused to confirm or deny the event, probably in an effort to further damage enemy morale — a standard policy for operations involving special forces.

SAS patrols can be inserted by various means, either by land, sea or air, for clandestine missions. Less often, the SAS adopt disguises and false identities for undercover operations. In Aden, for example, troopers took to wearing local dress to infiltrate the teeming markets that provided cover for local resistance groups. Under less unusual circumstances it is often possible for SAS teams to move into enemy territory from a 'friendly' area with relative ease, particularly if the borders are poorly monitored or consist of 'difficult' terrain. The fact that a number of journalists managed to get inside Iraq once the war was in progress points to the ease with which this method may be used by properly-trained SAS teams.

The SAS can now call upon a plethora of insertion and extraction methods

Clandestine insertion, entering enemy-held territory without his knowledge, is the most common method of infiltrating SAS four-man patrols. Insertion generally involves trekking across a hostile border, avoiding enemy patrols. This method was used during SAS cross-border operations during the Borneo Confrontation, but since that campaign technology has developed to such an extent that the SAS can now call upon a plethora of insertion and extraction methods involving parachutes, helicopters, submersibles and highly mobile 'go-anywhere' land vehicles.

Parachuting in a variety of forms is perhaps the best and most well-known insertion technique used by the SAS, and is one which places great emphasis on developing the relevant skills. Specially-trained SAS soldiers are capable of parachuting from aircraft flying at high-altitude, occasionally up to heights of almost 11km above sea-level, and each SAS 'Sabre' Squadron includes an Air or Freefall Troop consisting of 15 troopers, NCOs and their officer, all of whom are trained in high-altitude, low-opening (HALO) parachuting techniques.

The second most important method of parachute insertion is HAHO

HALO, one of two widely used parachute-insertion techniques, involves the patrol leaving an aircraft at high altitude, usually off the back ramp of a C-130 Hercules, and descending down as a group to a low altitude in freefall before deploying their 'chutes and landing

normally. The advantages of HALO are that aircraft flying so high are generally invisible to the naked eye, and that the team being inserted can form up and descend to a minimum height in a short time, thereby presenting less of a target to any unfriendly forces on the ground. Equally important is the fact that SAS teams inserted in this manner tend not to become separated. For almost all HALO descents an automatic parachute release mechanism, such as the new Irvin Hitefinder D/1 Mk 4, is used to deploy the parachute at around 760m. At this height the parachutist has a chance to deploy his reserve should the main canopy fail, and the patrol has time to select a suitable landing site. As most operational descents of this nature are conducted at night, the need for the patrol to remain together is clearly of great importance.

The second most important method of parachute insertion is HAHO (high-altitude, high-opening). For HAHO, exit procedures from the transport aircraft are fairly standard, but members of the patrol open their parachutes immediately after clearing the aircraft. In a typical HAHO descent the troopers would

Above: An example of the different jump altitudes of static-line and HALO parachuting. Note the breathing apparatus worn by the HALO soldier.

jump at around 10,000m, freefall for between eight to ten seconds, deploy their parachutes at about 8500m, and settle into a gentle gliding descent. Downwards speed will normally be around five to eight metres per second, depending on height, temperature, air density and other factors. A fully-laden SAS soldier parachuting in this manner from 10,000m using the GQ 360 main canopy can travel up to 24km before touching down. At these high altitudes oxygen is required for the initial part of the glide which, in total, can last well over an hour (70-80 mins).

The military benefits of the HAHO system for SF units are obvious. An aircraft can unload its human cargo outside the borders of a hostile country in comparative safety, yet the landing area can be many kilometres inside enemy territory. With the introduction of remotely-steered parachutes which can be fitted to stores containers, weapons and other military hardware can also be dropped with HAHO parachutists, who are able to guide in their equipment and land with it. Clearly, covert operations are more likely to succeed if both men and equipment land silently together without alerting any enemy forces in the vicinity of the drop zone.

Landing for both HAHO and HALO jumpers is a far safer affair than for static-line paratroopers. First, SAS freefall troopers can control their direction of drift and land into the wind, thereby ensuring a very compact drop, and second, they can control their rate of descent and stall their 'chutes on landing to reduce the likelihood of injury. The military freefall (MFF) parachutist can generally greet the ground standing and walk away, while the static-line parachutist, SAS or otherwise, must break his somewhat faster fall with a para-roll, and typically hits the ground like 'a sack of potatoes' — greatly increasing the chance of severe injury. The SAS, usually operating in small teams, cannot afford to lose a single operative at the very outset of a mission.

Below: The helicopter has greatly increased the tactical flexibility of special forces' missions as rotary-wing aircraft do not require landing sites. While the aircraft hovers SAS troopers can abseil down from the fuselage. In this way teams can be inserted into almost any type of terrain.

Another favoured method of airborne insertion employed by the SAS is by helicopter. Royal Navy (RN), Royal Air Force (RAF) and occasionally Army Air Corps (AAC) helicopters have provided the Regiment with the ability to move swiftly into the enemy's rear areas, often far beyond any recognisable front line, since they were first used to insert and extract SAS patrols in Malaya. Since the Regiment's offensive against the communist terrorists in Malaya, helicopter insertion has advanced a long way from the somewhat dangerous method of parachuting directly into the jungle canopy, known as tree-jumping, that was used in the campaign. This technique was a somewhat hit-and-miss affair and relied on the tree canopies to catch the parachutist, thus preventing him from hitting the floor of the jungle at speed. In many cases the parachutist's descent was not halted by the forest, with disastrous consequences for both the soldier and the operation. Altogether far too dangerous, the concept of tree-jumping and parachuting into dense jungle was dropped in favour of more reliable methods of clandestine insertion.

Helicopters are now in regular use by special forces throughout the world

Helicopters are now in regular use by special forces and airmobile units throughout the world, and the SAS has its own AAC flight based at its headquarters in Hereford. This is equipped with a variety of aircraft types including CH-47 Chinooks specially modified for long-range insertions with in-flight refuelling equipment. During the 1982 Falklands conflict the few helicopters not used to support the forthcoming ground campaign were tasked with supporting SAS reconnaissance patrols, inserting and extracting them throughout the islands and even farther afield. One RN Sea King dropped an SAS recce party on the Argentine mainland itself in a secret, but unsuccessful, mission to locate and destroy the Argentinian capability to launch Exocet missiles against the Task Force. Two more helicopters were lost

Above: The two-man Klepper canoe, made of wood and treated cotton, allows SAS units to silently infiltrate an enemy's coast and inland waterways.

during the ultimately successful attempt to recover a patrol trapped on the South Georgia's treacherous Fortuna Glacier prior to the main operations on the Falklands.

Helicopters are the most common method by which special forces can be inserted in-theatre by air. Current military tactical thinking has moved away from carrying vehicles and equipment beneath the fuselage, though this method is used over short distances. Over longer ranges the strain on the pilot is quite considerable and this, together with the reduction in the aircraft's overall range due to the increased drag, has resulted in special forces favouring helicopters which can accommodate vehicles in their fuselage. Aircraft that fulfil this requirement include the Sikorsky CH-53E Super Stallion and the Boeing Vertol CH-46 Sea Knight.

It can fit two dune buggy-type fast attack vehicles into its fuselage

The Super Stallion is an impressive aircraft, having the capability of carrying an internal payload of 31,638kg and having a top speed of 315km/hr. Its range of 2076km can be extended by means of its in-flight refuelling probe, and it can fit two dune buggy-type fast attack vehicles into its fuselage, enabling an SF team to go into action as soon as the helicopter has landed and the rear ramp is down (a major disadvantage of underslung vehicles is the additional time taken before the crew can go into action).

The Sea Knight is somewhat smaller than the Super Stallion but is, nevertheless, an extremely rugged helicopter which has been used by American SF units in Vietnam, Lebanon and Grenada. Only one specialist vehicle can be fitted into its fuselage and it is not equipped with an in-flight refuelling probe which limits its range.

The S-61 Sea King is one of the most successful helicopters ever built

Other rotary-wing aircraft suitable for clandesting missions include the Sea King and Puma helicopters. The Sikorsky S-61 Sea King is one of the most successful helicopters ever built. Having excellent range, speed and payload capacity — up to 15 fully-equipped soldiers can be accommodated within the fuselage — it is a superb utility aircraft. The Puma can carry 16 fully-equipped soldiers, has carbon composite rotor blades to substantially reduce maintenance and improve performance, and can also be fitted with 'polyvalent' air intake extensions to protect the engines when operating under icy conditions, thus making it suitable for SAS missions in an arctic environment.

Since the days when helicopters were used primarily as a means of extracting casualties that could not otherwise be reached, or were in need of particularly urgent medical treatment, to the recent hi-tech war in the Gulf, they have deployed and recovered countless SAS patrols throughout the world. Thanks to the introduction of passive night vision goggles, aircrew can pilot helicopters at extreme low-level at night and in low-light conditions, operational procedures that are much favoured by SAS teams wishing to be inserted or recovered from behind enemy lines. Other less technical innovations have been introduced over recent years, with the SAS often learning from other units and adopting or modifying their techniques. One such system is the STABO (Stabilizer, Tactical, Airborne Body Operations) harness used by US SF units and the Australian SAS Regiment in Vietnam. By clipping themselves onto single or multiple cables trailing below a helicopter, a four-man patrol can be lifted up and away without the need for an aircraft to land and load them aboard. STABO harnesses

Below: An SAS diver armed with an M16.

are similar to those used when parachuting, and can be carried in a bergen or worn as an integral part of the trooper's webbing. Suspended 18m beneath a helicopter travelling at 80km/hr or more 60m above the ground may not be everybody's idea of a good risk, but the STABO system is perhaps the best means of recovering patrols from a 'hot' landing site.

Apart from aerial methods of extraction and infiltration, seaborne methods are also used to insert SAS patrols. Each SAS squadron has its own amphibious specialists who belong to its Boat Troop. Trained in sub-aqua and surface swimming skills and the techniques required to use small surface craft and semi-submersibles, members of the Boat Troops have skills similar to those of the commandos of the Special Boat Squadron (SBS). SAS soldiers train with both open- and closed-circuit scuba systems, the latter preventing tell-tale bubbles from reach-

ing the surface to alert a watchful enemy; silent Klepper two-man canoes; high-speed Rigid Raider assault craft; and a variety of military inflatables and more conventional landing craft.

The SAS picked up many of the skills of Royal Marine Commando units

During the Borneo Confrontation the SAS picked up many of the skills of Royal Marine Commando amphibious units operating in the riverine areas, which used anything from local canoes to landing craft, and from these beginnings came the SAS's specialist Boat Troops. Over the years more sophisticated methods of inserting patrols from the sea have been developed, and seaborne operations played an important part in the Regiment's campaign during the Falklands conflict. The initial mission to retake South Georgia involved sea-

Above: Insertion via an enemy's coastline is a favoured special forces tactic. Units operating at night in small craft are invariably difficult to spot.

borne landings, undertaken at considerable risk in appalling weather conditions. Despite the fact that boats were overturned, flooded or suffered other damage, and one was swept out to sea, all the SAS personnel involved were eventually recovered once the island had been captured from the Argentinians. Clearly, the ability to deploy teams in the face of appalling sea conditions is of great advantage, not least in the area of surprise

Other amphibious operations involved parachuting troopers and equipment into the seas surrounding the islands, to then be picked up by submarines or surface vessels. Given the vastness of the areas involved and the sea temperatures, SAS operatives were equipped with special survival suits and homing beacons to facilitate the pick-ups. Patrols were then inserted by the boat close to the shore to begin their reconnaissance of the Argentinian positions prior to the main amphibious landings on West Falkland.

SAS patrols landing by sea along a hostile shoreline are usually inserted by Rigid Raider boats which are piloted by members of the most expert small-boat unit in the world: 539 Assault Squadron, Royal Marines. The boats themselves are five-metres long, have fibre-glass hulls and are fitted with 140hp Johnson outboard engines, or similar, and can reach speeds of up to 35 knots. Their low silhouette, coupled with modified, near-silent engines, make vessels such as these practical and effective craft for the clandestine infiltration and extraction of SF patrols along enemy-held coastlines. Tactically a troop of 16 SAS soldiers, including its officer, will usually be carried in three Rigid Raiders. On clandestine operations one of the craft would 'beach' on the shore first, with the other two in a position to provide covering fire if necessary. In addition, if one of the craft is disabled then its passengers and crew can be distributed between the other two vessels.

Below: A specially modified SAS Chinook helicopter at a base 'somewhere in England'. Note the in-flight refuelling probe at the front of the aircraft.

On at least one occasion a special operations LCVP had to return fire

Larger landing craft which, like the Rigid Raiders, were crewed by Royal Marine Commandos from the Task Force's Landing Craft, Assault and Raiding Squadrons, were also involved in the insertion and extraction of SAS and SBS reconnaissance patrols during the 1982 Falklands conflict. The large Landing Craft, Vehicle/Personnel (LCVPs) have a greater range (140 kilometres) but a slower speed (9.5 knots) than the more agile 'Raiders', but are more heavily armed and armoured. They can also be fitted with a range of light support weapons ranging from GPMGs to missile launchers, unlike the more vulnerable Rigid

Above: Re-creation of a reinforcement jump that occurred during the 1982 Falklands War. An SAS team and its equipment from Ascension Island were para-dropped into the South Atlantic from a Hercules transport aircraft and picked up by a Royal Navy ship.

Raiders, and on at least one occasion during the battle for the Falklands a special operations LCVP had to return fire.

These long-range patrol vehicles are based on the Land Rover chassis

Fixed-wing aircraft, helicopters, and sea-going raiding craft all play their role in helping the SAS soldier to reach his target, but wheeled vehicles similar to the heavily armed World War II SAS jeep still play an important part in inserting modern-day patrols into their operational areas. These long-range patrol vehicles, like the famed SAS 'Pink Panther' they are replacing, are based on the Land Rover chassis. These vehicles rely on speed and manoeuvrability to stay out of trouble and are essentially nothing more than modified long-wheelbase Land Rovers, stripped-down and mounting extra fuel tanks and weapons mounts. Other, lighter vehicles such as the Longline Light Strike Vehicle (LSV), recently used by the SAS Regiment in the Gulf War, have less range but are smaller and faster. The LSV is based on the civilian dune buggy and is driven by a powerful 1900cc engine, and many of its hydraulic, electrical and mechanical systems are duplicated so that it can sustain damage and still return to base. Getting personnel back safely is as important as getting them to the operational area, and the SAS Regiment prides itself on its ability to extract its patrols once their mission has been completed.

The development of Stealth technology — the value of which was proved in the 1991 war between Iraq and the United Nations — means that in the forseeable future SAS vehicles, fixed-wing aircraft, helicopters and amphibious vessels will be manufactured from extremely lightweight, composite radar-absorbent materials, thus reducing the likelihood of them being detected by the enemy. This will result in SAS insertion techniques being much more daring and novel than at present.

CONTROL
ZONE

BOOBY TRAPS

Vicious, uncivilised, simple yet deadly — booby traps are almost as old as war itself. Designed to maim rather than kill, to lower the morale of raw conscript enemy soldiers, and spread anger and fear among civilians, booby traps can be a very effective weapon in the hands of the elite soldier. Skill in the handling of explosives, together with an in-depth knowledge of the many traps of primitive peoples, means that an SAS operative can devise some savage yet simple devices in any type of environment.

Booby traps have long been used by both regular armies and guerrilla forces to capture an enemy unawares, and can be used either as part of a pre-planned ambush or left behind once a special forces (SF) patrol has left its area of operations. Booby traps are less conventional than the more traditional mines, but no less effective. Their immediate aim is often to create an air of uncertainty, fear and suspicion, or slow down the enemy by limiting his choice of movement, channelling his forces into areas where they can be engaged by other, more effective means. Booby traps force extreme caution on the part of the enemy, who have to expend considerable energy in

A bomb-damaged building in the Northern Ireland town of Inniskilling, the result of a large IRA explosive device. The Provisionals are highly skilled in making booby traps.

identifying and disarming these potentially lethal devices. Like many anti-personnel mines they are 'unconventional', often designed to injure and maim rather than kill, and as such have a severely detrimental effect on the morale of those unlucky enough to encounter them, as well as those who have to deal with the aftermath of a successful trap. Unlike the more traditional mines, booby traps can often be manufactured locally with as much variety and cunning as their makers wish to install in their deadly creations.

SF troops, such as the SAS Regiment or US Army Green Berets or Rangers, commonly operate in very remote areas and employ unconventional measures and techniques in their low-intensity operations. They need to know both how to construct and how to dismantle booby traps, often unsophisticated devices which, despite their relative simplicity, still play an extremely important part in modern-day hi-tech warfare. SAS training emphasises several aspects of booby-trapping: how to create booby traps from both local materials (wood, rock and vines, for example) and standard military equipment (mines, grenades and bullets, for example); how to position and conceal the devices for maximum effect; and, perhaps of greatest importance to SAS teams operating covertly, how to recognise and deal with the enemy's 'surprises'.

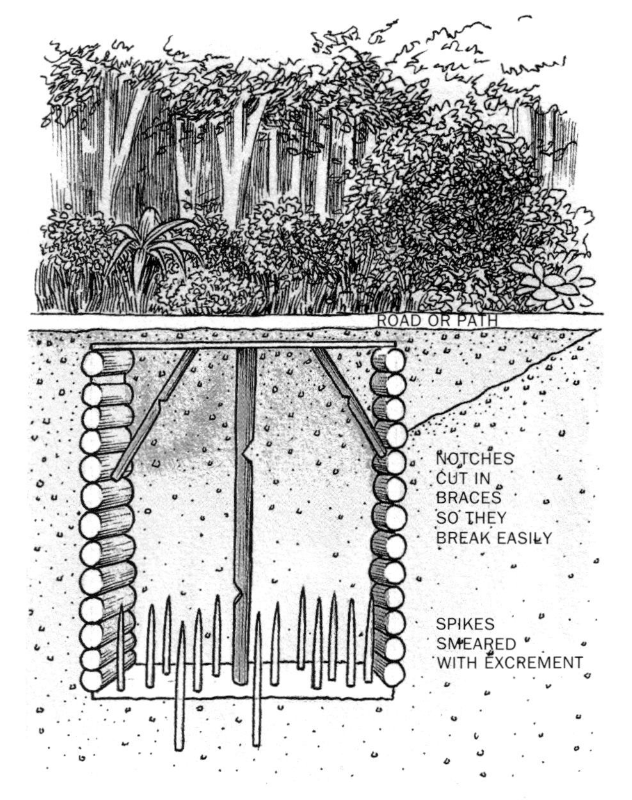

Above: To be effective pit traps must be camouflaged. Large traps can also damage vehicles.

Infiltration and exfiltration are important aspects of any SAS operation

Britain's SAS Regiment has spent decades developing the tactics, techniques and standard operating procedures (SOPs) that have led to it being classed as a world leader in small-scale operations with a strategic impact. Exploitation of an offensive operation, for instance immediately following an SAS attack on an enemy target, is one such tactic used by SAS units to augment their relatively small size. Because SF patrols operate in such small numbers it is vital that they make the maximum use of whatever resources they might possess or acquire.

One of the common dangers facing SAS fighting patrols is that their relatively small size makes them vulnerable to counterattack once the attack phase of a mission has been successfully completed and they have begun their withdrawal. Infiltration and exfiltration are both vitally important aspects of any SAS operation, but unlike infiltration, during which the patrol can rely on the advantages of concealment and ultimately surprise for its protection, exfiltration almost always takes place after the enemy has become aware of the patrol's presence in a particular area.

Ambush is one of the most common ways of dealing with enemy forces

In certain circumstances, for instance during clandestine cross-border SF operations similar to those practised by the SAS in Borneo and Indonesia, there is little or no likelihood of external support during the exfiltration phase. Traditional support, such as air-to-ground fire, artillery barrage and casualty evacuation (CasEvac), are all unavailable to personnel

involved in 'deniable' operations. So, too, is the provision by friendly forces of 'follow-up' troops to consolidate the objective, track down escaped enemy personnel, search the target area, and ambush any enemy reinforcements. In 'accountable' operations air support is most commonly used to extract SAS teams after they have completed their attack, but a patrol might find itself very much on its own until it has made its way back to friendly lines following a clandestine mission. It is under these circumstances that the booby trap or traps, if properly constructed and positioned, can be at its most devastatingly effective.

Above: A punji pit. Designed to impale a soldier unlucky enough to fall onto them, the stakes are covered with excrement to aggravate wounds.

Ambush is one of the most common ways of dealing with enemy forces following an SAS patrol once it has completed its attack, or has had its position compromised. Prior to its attack an SAS unit will undoubtedly reconnoitre one or more positions from which it can ambush pursuing enemy forces during its withdrawal phase. Likely trails or other routes that may be taken by enemy follow-up units will normally be targeted and plans laid for them to

Above: Crude but deadly. This IRA bomb would have been powerful enough to kill a British Army patrol travelling in an armoured personnel carrier.

be ambushed. There are two types of ambush that can be set during the withdrawal phase: manned and unmanned. The manned type rely on concentrated, heavy and accurate fire from a minimum number of troopers waiting in concealed positions. Unmanned ambushes have the same 'shock' value but do not rely on active troop involvement — usually devices are activated by an unwary enemy stumbling into them — to initiate the ambush, thus reducing the risk to members of the withdrawing force.

A track, ford or crossing point can be mined and/or booby trapped

A particular area such as a track, ford or crossing point can be mined and/or booby trapped so that the enemy force itself activates the ambush. One method is to use remote sensors, coupled with electrical or mechanical detonating systems. Remote ambushes have two major disadvantages: first, they might be sprung by friendly forces or a third party; second, they may become subject to electrical or mechanical failure. However, if they are well-positioned by trained specialists, most modern systems, such as those employed by the SAS, are unlikely to detonate by mistake. If there is a risk of the ambush being activated by local civilians, for example, one or more members of the patrol will normally stay behind to spring the ambush by command detonation.

The specialist knowledge of the individual SAS trooper in how to site ambushes, and how to construct and position booby traps and mines, is useful both in preventing him from becoming a casualty, and for him to inflict physical and psychological injury on his enemies. Although it is standard practice for SAS patrols to avoid areas where there is a danger of encountering enemy-positioned ambushes or booby traps, there are occasions when such places cannot be avoided. The knowledge of where the enemy might position his deadly surprises helps the SAS soldier decide the best place to conceal his own.

Buildings and abandoned equipment almost always attract the unwary

Possible ambush sites such as bottlenecks on routes and tracks, crossing points, obstacles, ditches, defiles and natural cover are all areas where mines and booby traps can be used to their maximum effect. Buildings, shelters and abandoned equipment almost always attract the unwary, tired soldiers or guerrillas, and are favourite positions for traps or mines.

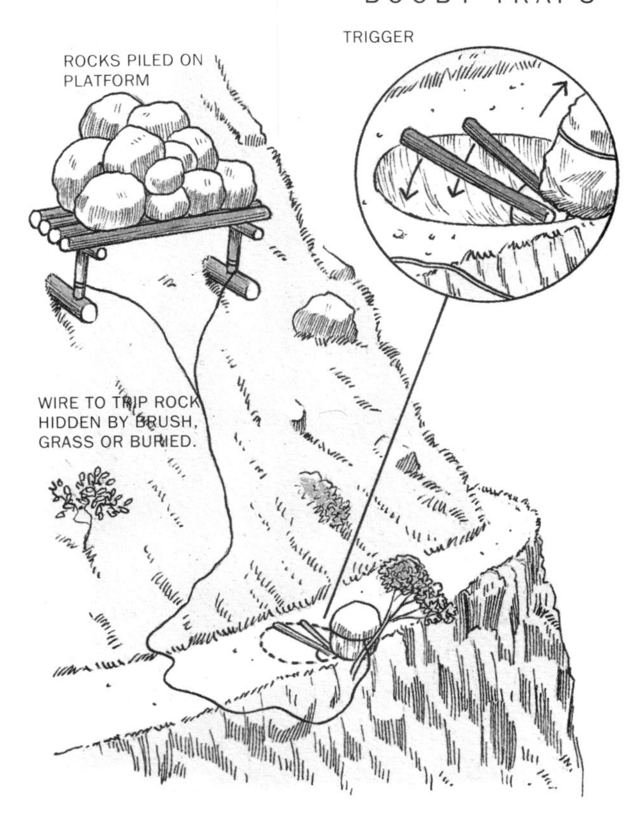

Right: Diagrammatic representation of a rockfall trap. In reality, the rocks, platform and wires would be heavily camouflaged so as not to alert an enemy patrol. The platforms themselves can be built to hold a 5000kg load of rocks!
Below: A selection of home-made Viet Cong booby traps. From left to right: a grenade made from a ration can; a block of TNT wrapped with nails; a can with a home-made fuze; and a cloth bag filled with plastic explosive, nails and broken glass.

Above: The heart of the Jivaro Catapult, employed by the Amazonian Jivaro tribe, is the tightly-wound spring band that holds and swings the spiked paddle. The trap can inflict horrific injuries.

For conscript troops the first indication as to the presence of concealed explosives, booby traps or mines is when one or more fall victim to such a device. The best defence against these threats is the almost instinctive recognition of all factors associated with them. Experience, training, visual alertness and general awareness are all important qualities and skills in identifying and locating such threats.

There is rarely time to completely conceal mines and booby traps

Whether recognising likely ambush positions or placing them himself, the SAS soldier needs to know the signs that may indicate their presence. There is very rarely time to completely conceal mines and booby traps, and some indication, such as earth spoil or the disturbance of vegetation, usually gives a clue to the observant SAS trooper or well-trained enemy. Particular patterns left in the earth by footwear, vehicle tyres or tracks can suggest the placing of mines, as they indicate a well-used route that would be a prime candidate for booby-trapping. Flattened or loose earth both indicate an attempt to conceal or camouflage something underneath, and a partial obstacle across a route or track can mean an attempt to redirect the traveller into danger. Trip wires are suspicious in any event and indicate either mines, booby traps or an attempted deception.

Both enemy and friendly minefields pose a threat to the SAS soldier, who may have to negotiate both types in the course of his mission, possibly on the outward leg as well as the return. Minefields sown with conventional anti-personnel and anti-armour mines are best avoided, as they can severely slow down a patrol and would normally have to be negotiated at night — a risky business.

Modern mines are usually lethal. A typical example of an anti-personnel mine is the ARGES Bounding Mine produced by Austria. It is a large cylindrical device weighing 6kg which is mounted on a base-plate, and which has three trip wires stretching radially from the

Above: The Chinese Chopper requires perfect balance if the log is to fall on top of the target.

mine. A soldier hitting one of these wires causes the firing pin to release a tensioned spring, sending the mine shooting into the air. The mine and the base-plate remain in contact by means of a thin 1.5m-long wire. As the mine is propelled into the air the wire is stretched taut, thus detonating the charge. The ensuing explosion sends 4600 metal balls outwards at high velocity at approximately chest height. In common with modern military thinking this mine is not designed to kill but to wound, the idea being that a wounded man is more of a liability to the enemy than a dead one.

Mechanical minelayers have reduced the time needed to sow a minefield

One of the consequences of modern mechanised warfare has been the improvements in minelaying techniques, which means that mine fields are now larger and more dangerous than ever. This can be extremely time consuming for SF teams who have to penetrate them. Mechanical minelayers have significantly reduced the time needed to sow a modern minefield. The new American Ground-Emplaced Mine-Scattering System (GEMSS), for example, can lay a 2500m surface minefield in around six hours. Soviet mine fields are traditionally wide and very dense, making their penetration by SF units extremely difficult. Most of their anti-personnel mines have been designed to be inserted into the ground by a number of simple methods. One such method involves dropping the mines down chutes attached to the rear of trucks and which trail on the ground. The mines slide down the chutes and are armed and concealed by the men following the truck. Though this may seem slow, the system can lay over 200 mines in under 20 minutes.

The device should be examined without disturbing the mechanism

Unless a clear route across the minefield could be found, and if there were such a route it would no doubt be covered by artillery or support fire if not actually guarded, the mines would have to be located by hand and by-

Right: A mortar shell, lodged loosely in a tree, will crash down when troops below knock the tree.

the device until it detonates, or by destroying it in place with a small amount of explosive. Both these methods mean that nothing can be gained from the experience, no data on the device's mechanisms, for example, but it is the safest method in most circumstances. However, by temporarily neutralising the trap or mine the device can be adapted to suit the purposes of the de-fuzer. In the past SAS units that have de-activated enemy booby traps have been known to re-activate them after modification, much to the surprise of those who laid them in the first place. Just as animal traps are visited by the trapper to check the catch, booby traps are often visited by their creators to test the system or readjust its position.

Another hazard is the trip wire, which may be used to detonate any mine

Another hazard is the trip wire, which may be used to detonate any type of mine. It is commonly used with the small, hard-to-detect anti-personnel variety; frequently these are mines which spring into the air before exploding. If positioned correctly, often parallel to the enemy's line of march or above ground level, trip wire-detonated mines or specially modified grenades can cause horrific casualties. Trip wires are usually made from non-reflective, dull metal, normally painted matt green or brown, and are difficult to spot when suspended at ankle height amongst tall grass. In forest or jungle areas, for example, they may be used in conjunction with anti-personnel mines or other booby traps suspended high in the foliage for the most devastating effect on troops passing underneath. Like mines concealed beneath the ground, trip wires can best be located by touch and, instead of working with a prodder such as a bayonet or similar, a feeler made from stiff wire should be used. Re-shaped metal coat hangers make useful feelers, and held between thumb and forefinger should be swung to and

Left: The mountain trail trap. The leading soldier kicks away the trigger causing the log to roll over the edge, taking the whole patrol with it.

passed. Detection by electronic (mine-detectors or chemical 'sniffers') or canine (specially-trained dogs) methods is outside the normal scope of SAS operations and is best left to the experts, in the case of the British Army sappers from the Royal Engineers. SAS soldiers have to rely on more traditional methods. Prodding, although slow and laborious, is the only effective means of detecting all types of ground-sown mine, but is not the safest as a prodded mine may detonate, wounding or killing those men in the immediate vicinity.

By temporarily neutralising the trap or mine the device can be adapted

Once located, mines and traps should be neutralised unless the aim of the patrol is to pass through an obstacle, such as a minefield, as quickly as possible. In this particular case the mines and wires would normally be marked and the remainder of the patrol would then move around them. The modern practice of booby-trapping mines by fitting them with anti-lift devices has made neutralising and removing them a dangerous business. If they must be dealt with there are some basic ground rules that must be adhered to. The device should be examined thoroughly without disturbing the mechanism. Protruding wires, whether taut or slack, should not be cut until the ends have been traced, and while doing so should not be disturbed in any way. Some devices are detonated when ultra-thin metal wires are broken. Once the device has been checked and the operator is convinced he understands the mechanism, it should be safe to remove the fuze and the leads between it and the main charge, being aware all the time that a release switch, or similar, may have been attached to the explosive.

Apart from clearing a trap or mine by removing the fuze and lifting it out, it can be detonated remotely. This can be done either by employing a cable and grapnel, and dragging

Below: Two US soldiers in Vietnam assist a comrade who has just stepped on a small spike trap, both injuring him and demoralising his unit.

fro over the area to be checked.

The laying, detecting and clearing of booby traps requires a great deal of theoretical knowledge of different systems, as well as practical experience. Such tasks are usually the responsibility of trained combat engineers but, because SAS squadrons often operate unsupported, the burden of laying and clearing booby traps occasionally falls on the individual SAS soldier. In low-intensity operations, especially in a jungle environment such as Malaya or Borneo, local inhabitants may be able to provide information as to the location of enemy mines, ambushes and traps. Innocent civilians often suffer as a result of indiscriminate placement of booby traps, and may be only too willing to help the SAS by pointing out such threats.

Another aspect of booby traps is the use of the 'home-made' variety

Another aspect of booby traps is the use of the 'home-made' variety. One consequence of the Regiment's many years of service in the jungles of the Far East has been the gaining of much valuable knowledge concerning the construction of large 'animal' traps. During in-theatre survival training all troopers are instructed in the techniques required to construct bird, monkey and large animal traps for the purposes of supplementing their diet. However, some of these traps can be used with little modification to kill or maim a more deadly and dangerous animal: man. The majority of such devices rely on the effect of gravity or suddenly-released tension in their operation. Pig traps, especially those that propel a sapling-mounted spear connected to a trip wire, can seriously injure or even kill a man as well as an animal.

Special forces teams employ a combination of old and new

While modified animal traps can be used against a human enemy, some are specifically manufactured with man in mind. Many valuable lessons have been learned from the Viet Cong for instance, whose long battle against the Japanese, French and American armies led to the modification of animal traps and the incorporation of captured or discarded modern munitions in a variety of booby traps. Punji sticks, slivers of metal or bamboo sharpened and smeared with faecal matter, were set into the ground or in dead-fall traps and then camouflaged. The weight of a passing infantryman was sufficient to propel him through the flimsy camouflage cover and impale his foot on the sticks. Unlikely to kill an individual, these devices nevertheless produced severe infections that took a long time to heal, and certainly had an unnerving effect on those who witnessed their effects. Adapted high-explosive rounds, or individual rounds of small-arms ammunition, linked-up to a locally-made pressure plates worked on a similar basis, but with much more lethal consequences.

On occasion SF teams, especially those operating in remote inaccessible rural areas, employ a combination of old and new. Modern

Below: Spike traps should be placed at the bottom of a small incline or at the top of a hill, places where the targets — horses or men — are likely to step down harder than normal.

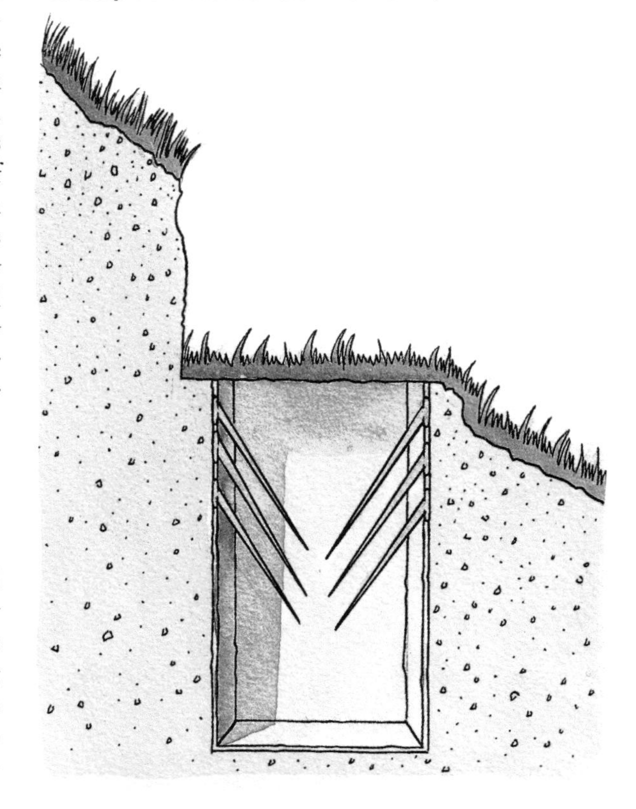

anti-personnel Claymore mines (directional devices comprising hundreds of steel balls backed by an explosive charge that can be detonated by command or trip wire) of the type used in many SAS ambushes are also very effective in defence, and have been employed by British and US special forces to secure their isolated bases in Malaya, Borneo and Vietnam.

It is no use disguising a trap with bushes if none grow there naturally

What are the ground rules when setting traps? The guiding principle is that any booby trap must be in harmony with its surroundings; every detail concerning its concealment and blending in must be addressed, otherwise it will not work. If the trap involves rolling logs down a hill, for example, there must be logs occurring naturally in the immediate area. Similarly, it is no use disguising a trap with bushes if none grow there naturally, nor can pits be dug in swamps or on the tops of rocky mountains. Attention must also be given to even the most insignificant details; ropes or trip wires, for instance, should not be covered with limbs from trees that do not grow in the immediate vicinity of the trap.

When a trap is being constructed the surrounding terrain must be altered as little as possible, this can often mean that the trap's mechanisms will have to be built well away from the area in which it is to be set. When the booby trap is in place any dirt, rocks or sticks that have been unearthed in the process must be moved away and hidden. For individual SAS soldiers this work is tedious, labour intensive and slow. The work should be undertaken by as few patrol members as possible, more people involved just creates greater confusion and causes increased scaring of the ground with signs of activity, thus increasing the difficulties of concealing the trap.

Above: An IRA car bomb. These devices, lightweight and difficult to spot, can be detonated by remote control or by means of a timer. They are one of the most effective terrorist booby traps.

After the trap has been set the area must look natural

After the trap has been set the area must look natural — no unusual ground or foliage disruption, no wood shavings or newly-scraped rocks. Designers of traps have often gone to incredible lengths to conceal their work. In Africa, for example, warriors have been known to shape logs to make them resemble animal tracks and place them on poles, which were then 'walked'

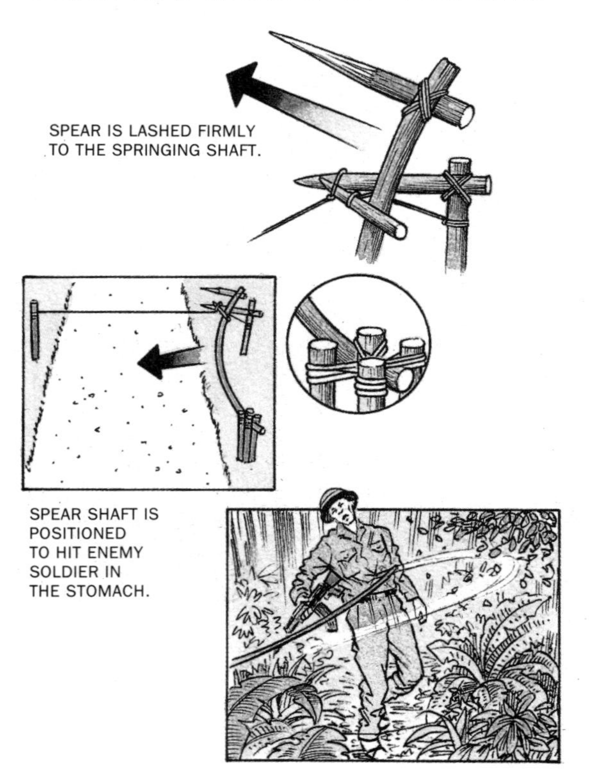

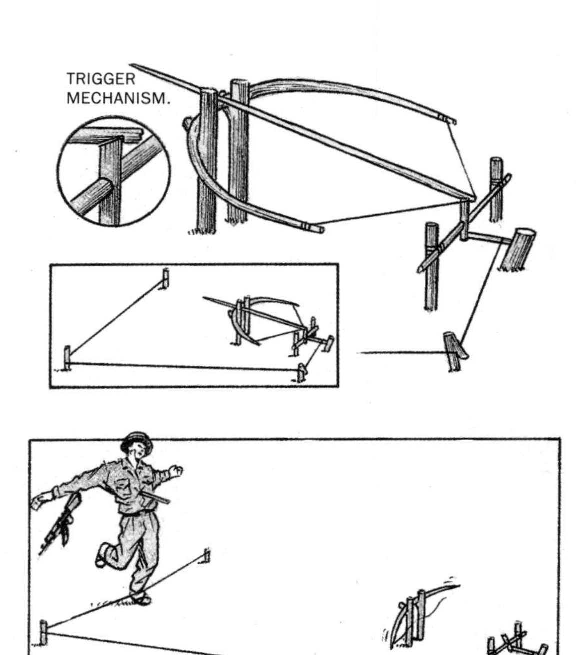

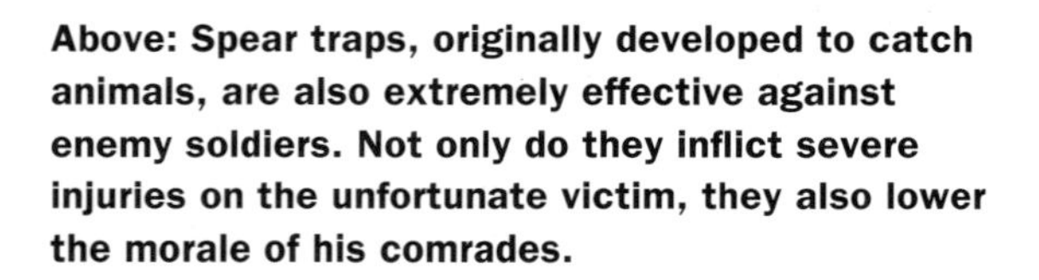

Above: Spear traps, originally developed to catch animals, are also extremely effective against enemy soldiers. Not only do they inflict severe injuries on the unfortunate victim, they also lower the morale of his comrades.

through the booby-trapped track. As the tracks looked extremely natural they convinced the enemy to walk into the trap. This method can even deceive highly experienced enemy scouts, who will know that wild animals are more cautious of where they walk than people are.

Catching the right quarry is extremely important

SAS teams setting booby traps will always be alert to the dangers that the traps can pose to innocent civilians. Catching the right quarry is extremely important. A small native child impaled on punji stakes at the bottom of a pit trap can have a detrimental affect on the 'hearts and minds' battle to say the least. To guard against this SAS patrols, after having won over the local inhabitants to their cause, will warn them of any traps in the area. The only real danger is to very small children, who are extremely willful and can sometimes wander long distances from their villages.

The grenade is then left on a roof, or in a gutter or sewer

The urban environment presents the SAS trooper with a different set of challenges with regard to booby traps, though his work can have just as destructive results. The emphasis in towns and cities is on using man-made materials in what is, after all, a man-made environment. A simple hand grenade, for example, can be a particularly useful tool to have when setting traps. One can be placed in a tin of wet mud with the top of the grenade and the pin let exposed. After the mud has dried the grenade, complete with caked mud wraparound, is removed from the can and the pin removed. The grenade is then left on a roof, or in a gutter or sewer. After several rainstorms the mud washes off, thus releasing the arming lever and activating the grenade. This could take

place weeks after it was emplaced. Alternatively, a grenade can have its arming lever secured by tape wrapped around the body. The pin is pulled and the grenade is left in a pile of rubbish which is to be burned. When the fire is set the tape will be burned away, thus activating the grenade.

When the light switch is flicked on the explosive detonates

Inside buildings light receptacles can be used to cause death and destruction. A light bulb is removed from its socket and an explosive suspended near the receptacle. An electric blasting cap is then inserted into the charge and the wires from the cap are attached to the inside of the light socket itself. When the light switch is flicked on the explosive detonates, sending blast waves throughout the building.

Another booby trap can be constructed from household materials. A piece of wire is fixed to both the upper and lower lips of a clothes peg. When the clothes peg is closed the bare wires must make contact with each other. A piece of wood or similar non-conducting material is placed between the lips of the clothes peg to ensure they do not touch. The

Above: This Royal Ulster Constabulary Land Rover was destroyed by an IRA car bomb. Though the bombs themselves are relatively simple, the terrorists in Northern Ireland have access to extremely sophisticated detonation systems.

peg is secured against a post, tree or piece of furniture. A trip wire is then fixed in place with one end attached to the non-conducting material in the clothes peg. The wires fitted to the lips of the clothes peg are attached to a blasting cap, a battery and the explosive. When the trip wire is hit, the wooden peg is knocked out from between the lips of the clothes peg, allowing both bare wires to touch, thus forming a circuit which sets off the explosive.

Such diversions allow SAS patrols to operate unmolested

Booby traps set in an urban environment can cause death and destruction out of all proportion to their size, and they can also cause widespread fear and demoralisation. For the enemy sweeping areas to ensure they are clear of booby traps can be time consuming and can tie down large numbers of troops. Such diversions allow SAS patrols to operate unmolested,

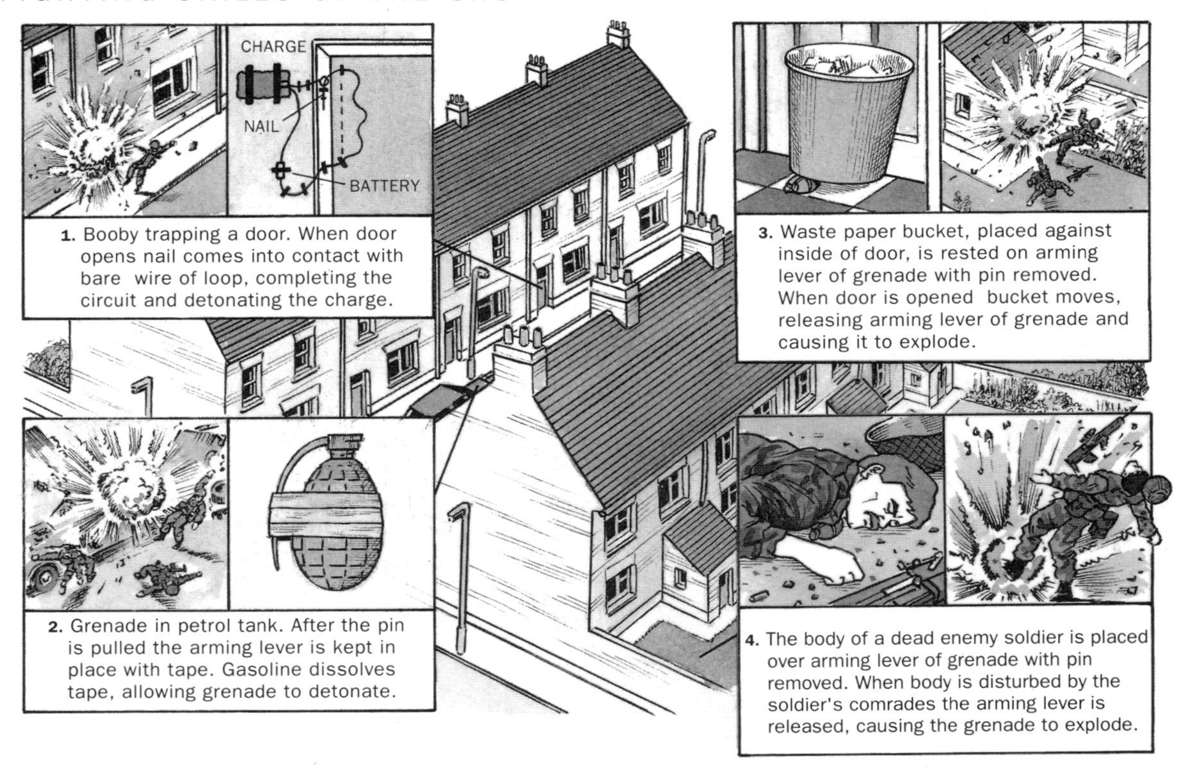

Above: Various booby traps that can be set in towns and cities. With a little imagination and the right materials the urban environment can be turned into a very hostile theatre of operations.

free to carry out their mission or, alternatively, withdraw safely after an operation.

The average modern soldier is ill-prepared for booby traps

It should not be assumed that the primary role of SAS soldiers on operations is the laying of urban booby traps, as this would make them more like ruthless terrorists that elite soldiers. Nevertheless, an in-depth knowledge of the materials and construction of traps and self-detonating devices can be a bonus to a patrol. The creation of chaos and uncertainty acts as an effective smoke screen behind which SAS units can carry out their missions.

Its is a simple fact that the average modern soldier is ill-prepared for booby traps. This may seem surprising given the thorough training that they receive. Nevertheless, the experience of soldiers and policemen in Ulster is testimony to the fact that simple home-made devices can often be the most effective. This also holds true in other environments. Soldiers on patrol in the jungle, for example do not expect to run into a falling log or a springing tree limb as they are usually too busy looking out for enemy soldiers. The SAS soldier will be aware of this and can thus set his traps with confidence. He, on the other hand, will have an eye for the unusual and will notice every detail around him.

Britain's SAS has never been above using tried and tested booby traps

Larger anti-tank and vicious anti-personnel mines, medieval-like oil pits, spiked mantraps, razor wire and hand-sharpened wooden spikes have all been used together to protect special operations bases in the Far East and elsewhere. The moral is 'if it works — use it' and, with its incomparable experience of counter-insurgency missions and guerrilla warfare, Britain's SAS has never been above using tried and tested booby traps, no matter how old they are, and regardless of their country of origin, in its countless operations.

GLOSSARY

WEAPONS AND EQUIPMENT USED BY THE SAS

Accuracy International PM
Type: bolt-action sniper rifle
Calibre: 7.62mm
Magazine capacity: 10 rounds
Range: 1000m
Rate of fire: single shot
Weight: 6.5kg
Length: 1124mm

Browning High Power
Type: semi-automatic handgun
Calibre: 9mm
Magazine capacity: 13 rounds
Range: 40m
Rate of fire: single shot
Weight: 0.927kg
Length: 190mm

C-130 Hercules
Type: tactical transport aircraft
Maximum speed: 618km/hr
Range: 4002km
Crew: four/five
Armament: none
Passengers: 92 troops

CH-47 Chinook
Type: transport helicopter
Maximum speed: 304 km/hr
Range: 304km
Crew: three
Armament: none
Passengers: 44 troops

Colt Commando
Type: assault rifle
Calibre: 5.56mm
Magazine capacity: 30 rounds
Range: 200m
Rate of fire: 700-800rpm (cyclic)
Weight: 3.23kg
Length: 787mm

G3
Type: assault rifle
Calibre: 7.62mm
Magazine capacity: 20 rounds
Range: 600m
Rate of fire: 500-600 rpm (cyclic)
Weight: 5.54kg
Length: 1025mm

HK53
Type: submachine gun
Calibre: 5.56mm
Magazine capacity: 40 rounds
Range: 400m
Rate of fire: 600rpm (cyclic)
Weight: 3.68kg
Length: 765mm

Ingram Model 10
Type: submachine pistol
Calibre: 9mm
Magazine capacity: 30 rounds
Range: 40m
Rate of fire: 1145rpm (cyclic)
Weight: 2.84kg
Length: 548mm

Ingram Model 11
Type: submachine pistol
Calibre: 9mm
Magazine capacity: 30 rounds
Range: 40m
Rate of fire: 1200rpm (cyclic)
Weight: 2.84kg
Length: 460mm

Javelin
Type: man-portable surface-to-air missle (SAM) system
Range: 5km
Maximum altitude: 2km
Length: 1.4m

L2A3 Sterling
Type: submachine gun
Calibre: 9mm
Magazine capacity: 34 rounds
Range: 200m
Rate of fire: 550rpm (cyclic)
Weight: 3.5kg
Length: 864mm

L34A1 Sterling
Type: silenced submachine gun
Calibre: 9mm
Magazine capacity: 34 rounds
Range: 150m
Rate of fire: 515rpm (cyclic)
Weight: 4.3kg
Length: 864mm

L7A2
Type: general purpose machine gun
Calibre: 7.62mm
Magazine capacity: belt feed
Range: 1800m
Rate of fire: 750-1000rpm (cyclic)
Weight: 10.9kg
Length: 1232mm

Land Rover
Type: long-range reconnaissance vehicle
Maximum speed: 120km/hr
Range: 800km
Crew: three
Armament: two 7.62mm general purpose machine guns; one at the front, the other mounted at the rear. Also fitted with front and rear smoke dispensers
Passengers: none

Law 80
Type: anti-tank weapon
Calibre: 94mm
Magazine capacity: one shot throw-away
Range: 500m
Rate of fire: single shot
Weight: 9.5kg
Length: 1.5m

Light Strike Vehicle
Type: high mobility vehicle for operations in hostile environments
Maximum speed: 100km/hr
Range: 200km
Crew: two
Armament: Vehicle can be armed with a variety of weapons including 7.62mm general purpose machine gun, .50 calibre machine gun, twin Browning machine gun, 40mm M19 grenade launcher, 30mm cannon, Milan missile, 60mm mortar and .50 calibre Gatling gun
Passengers: none, though 12 troops can be carried in emergencies

M16A2
Type: assault rifle
Calibre: 5.56mm
Magazine capacity: 30 rounds
Range: 500m
Rate of fire: 950rpm (cyclic)
Weight: 3.72kg
Length: 990mm

M203
Type: grenade launcher
Calibre: 40mm
Magazine capacity: no magazine on weapon
Range: 350m
Rate of fire: single shot
Weight: 1.63kg
Length: 394mm

Milan
Type: anti-tank weapon
Calibre: 90mm
Magazine capacity: no magazine on weapon
Range: 2000m
Rate of fire: 4rpm
Weight: 23.08kg
Length: 770mm

MP5A3
Type: submachine gun
Calibre: 9mm
Magazine capacity: 30 rounds
Range: 200m
Rate of fire: 800rpm (cyclic)
Weight: 2.88kg
Length: 680mm

MP5K
Type: submachine gun
Calibre: 9mm
Magazine capacity: 30 rounds
Range: 200m
Rate of fire: 900rpm (cyclic)
Weight: 2kg
Length: 325mm

MP5SD
Type: silenced submachine gun
Calibre: 9mm
Magazine capacity: 30 rounds
Range: 200m
Rate of fire: 800rpm (cyclic)
Weight: 3.1kg
Length: 780mm

PRC 319
Type: lightweight radio system
Number of channels: 20
Frequency range: 1.5 - 40MHz
Battery life: 500 hours (standby mode)
Weight: 3.4kg
Protection: system is able to withstand salt contamination, dust, sand, rain and nuclear effects

SA-80
Type: assault rifle
Calibre: 5.56mm
Magazine capacity: 30 rounds
Range: 400m
Rate of fire: 800rpm (cyclic)
Weight: 4.68kg

SA 330 Puma
Type: tactical transport helicopter
Maximum speed: 293km/hr
Range: 550km
Crew: two
Armament: none
Passengers: 20

Self-Loading Rifle (SLR)
Type: assault rifle
Calibre: 7.62mm
Magazine capacity: 20 rounds
Range: 600m
Rate of fire: single shot
Weight: 5.07kg
Length: 1430mm

Stinger
Type: man-portable surface-to-air missile (SAM)
Calibre: 70mm
Flight speed: Mach 2
Range: 5km
Rate of fire: single shot
Weight: 13.4kg
Length: 1524mm

Tube-launched, Optically-tracked, Wire-guided (TOW) missile
Type: anti-tank missile
Calibre: 152mm
Flight speed: 278m/sec
Range: 3.75km
Rate of fire: single shot
Weight: 28.1kg
Length: 1.16m

NOTES
RPM: rounds per minute.
Cyclic rate of fire: the theoretical rate of fire possible with an infinite and continuous supply of ammunition.

PICTURE CREDITS

Accuracy International 24 **Phil Asher** 164 **Aviation Photographs International/Flack** 25, 32, 35, 40, 60, 61, 69, 83, 119, 129, 135, 137, 146, 167, 174 **Aviation Photographs International/Shakespeare** 6, 93, 151 **Brown Packaging** 17, 21, 49, 187 **Camera Press** 127, 179, 181, 184, 185 **Ian Hogg** 15 **Imperial War Museum** 74, 78, 81, 116, 142, 152, 157 **PMA Pictures** 9, 11, 173 **Hugh McManners** 63, 105 **The Military Picture Library** 13, 91, 141, 155, 161, 165 **Pacemaker** 45, 65, 67, 97, 177, 180, 189 **Photographer's International** 27 **Press Association** 51 **Rex Features** 59 **SAS Regimental Association** 33, 71 **Jim Shortt** 46, 57 **USMC** 85 **Soldier Magazine** 30 **Syndication International** 37 **Thorn EMI** 107 **Telegraph Colour Library** 75, 159, **TRH Pictures** 19, 43, 101, 113, 117, 123, 145, 163 **Valpolini** 54 **Bernard & Graefe Verlag** 53